Lecture Notes in Computer Science 3028

Commenced Publication in 1973
Founding and Former Series Editors:
Gerhard Goos, Juris Hartmanis, and Jan van Leeuwen

T0219843

Springer
*Berlin
Heidelberg
New York
Hong Kong
London
Milan
Paris
Tokyo*

Daniel Neuenschwander

Probabilistic and Statistical Methods in Cryptology

An Introduction by Selected Topics

 Springer

Author

Daniel Neuenschwander
Universities of Bern and Lausanne (Switzerland) and
Swiss Ministry of Defense
Section of Cryptology
3003 Bern, Switzerland
E-mail: daniel.neuenschwander@bluewin.ch

Library of Congress Control Number: 2004105111

CR Subject Classification (1998): E.3, G.3

ISSN 0302-9743
ISBN 3-540-22001-1 Springer-Verlag Berlin Heidelberg New York

Springer-Verlag is a part of Springer Science+Business Media

springeronline.com

© Springer-Verlag Berlin Heidelberg 2004
Printed in Germany

Typesetting: Camera-ready by author, data conversion by Boller Mediendesign
Printed on acid-free paper   SPIN: 10998649   06/3142   5 4 3 2 1 0

To Galina

Preface

Cryptology is nowadays one of the most important subjects of applied mathematics. Not only the task of keeping information secret is important, but also the problems of integrity and of authenticity, i.e., one wants to avoid that an adversary can change the message into a fraudulent one without the receiver noticing it, and on the other hand the receiver of a message should be able to be sure that the latter has really been sent by the authorized person (electronic signature). A big impetus on modern cryptology was the invention of so-called public-key cryptosystems in the 1970's by Diffie, Hellman, Rivest, Shamir, Adleman, and others. In particular in this context, deep methods from number theory and algebra began to play a decisive role. This aspect of cryptology is explained in, for example, the monograph "Algebraic Aspects of Cryptography" by Koblitz (1999). The goal of these notes was to write a treatment focusing rather on the stochastic (i.e., probabilistic and statistical) aspects of cryptology. As this direction also consists of a huge literature, only some glimpses can be given, and by no means are we always at the frontier of the current research. The book is rather intended as an invitation for students, researchers, and practitioners to study certain subjects further. We have tried to be as self-contained as reasonably possible, however we suppose that the reader is familiar with some fundamental notions of probability and statistics. It is our hope that we have been able to communicate the fascination of the subject and we would be delighted if the book encouraged further theoretical and practical research.

Let me give my gratitude to my colleagues in the Cryptology Section in the Ministry of Defense of Switzerland for the excellent and stimulating working atmosphere. Many thanks are also due to Werner Schindler from the German "Bundesamt für Sicherheit in der Informationstechnik" for helpful discussions. Furthermore, I am indebted to Springer-Verlag, Heidelberg for the agreeable cooperation. However, the most important thanks goes to my wife Galina for her constant moral support of my scientific activities. Without her asking "How is your book?" from time to time, the latter would certainly not yet be finished!

Bern, February 2004   Daniel Neuenschwander

Contents

Introduction

Background

Cryptology is nowadays considered as one of the most important fields of applied mathematics. Also, aspects from physics and, of course, engineering science play important roles. Classical cryptology consisted almost entirely of the problem of secret keeping. The so-called "Caesar shift code" was just a shift of the alphabet by a certain number of places, e.g., 3 places (then the plaintextletter "a" was encrypted by the ciphertextletter "D", "b" by "E", etc., "w" by "Z", and then "x" by "A", "y" by "B", "z" by "C"). Such a shift code is, of course, trivial to decrypt[1], because one needs to try only 25 possibilities with some groups of subsequent ciphertextletters until one obtains some meaningful plaintext. More general are monoalphabetic substitutions, which are just any permutation of the alphabet. Here, one has $26! - 1 \approx 4 \cdot 10^{26}$ possibilities, but as the same plaintextletter always corresponds to the same ciphertextletter and vice versa, frequent letters (or pairs/triples of letters) in the ciphertext will with great probability correspond to frequently occurring letters (pairs/triples) in the language in which the plaintext is written, for example the letter "e" in German. For example, the following features of German language support the decryption of monoalphabetic encryptions: If in the ciphertext a triple of consecutive letters occurs several times, then there is a good chance that it corresponds to the plaintext triple "sch"; the plaintext letter "c" is almost always succeeded by "h" or "k", "q" by "u" with hardly any exceptions. In any language (and also with more general cryptosystems) the encryptor should avoid the use of "mots probables" (words from which an adversary can conjecture that they appear in the plaintext, e.g., military terms, "Heil Hitler", etc.). During the Second World War, this danger was often neglected, a mistake that was not the most important, but one of several reasons why enemy codes were decrypted in a decisive measure at that time. In recent years, many documents have been (and still are) found by historians in archives which confirm this fact. In the year 1586, the French diplomat Blaise de Vigenère (1523-1596) found a polyalphabetic code that

---

[1] In all our subsequent text, the word "decipher" will mean the decoding of a ciphertext by its legitimate receiver, whereas "decrypt" will mean the breaking of the code by an adversary.

D. Neuenschwander: Prob. and Stat. Methods in Cryptology, LNCS 3028, pp. 1-7, 2004.
© Springer-Verlag Berlin Heidelberg 2004

was thought to be "unbreakable" for centuries. This code will be presented in Section 1.1 of our text, together with the attacks on it found not earlier than in the second half of the 19th and at the beginning of the 20th century. After the spectacular successes in decrypting rotor enciphering machines such as ENIGMA, etc., during the Second World War, in the second half of the 1970s a great impetus on the development of modern cryptology was given by the invention of so-called public-key cryptosystems, in particular the code that is now known under the name "RSA system" (named after the authors who published it, namely "R" for Rivest, "S" for Shamir, and "A" for Adleman). Its detailed working is described in Section 2.1. The only nontrivial ingredient is Fermat's Little Theorem, which was known as a piece of "pure" number theory long before. It turned out since then that number theory and algebra are of decisive importance in modern cryptology, both in cryptography and cryptanalysis, in contrast to the assertion of the English mathematician G. Hardy (1877-1947) that by analyzing primes one "can not win wars"!

Nowadays, not only (classical) algebra and number theory, but also many other fields of mathematics, such as highly advanced topics of algebra and number theory (such as, for example, modern algebraic geometry, elliptic curves), graph theory, finite geometry (see, for example, Walther (1999)), probability, statistics, etc., play a role in cryptography, not to mention the recent (at least theoretical) developments in quantum computing and quantum cryptography (based on quantum mechanics) and all questions on hardware implementation of cryptosystems.

Furthermore, other goals entered into cryptology, namely the task of securization of the integrity and authenticity of a message. This means that (even for a possibly open transmission channel) one wants to avoid the message being changed by some unauthorized person without the receiver noticing it, and, on the other hand, the receiver wants to be sure that really the authorized person was the sender of the message (electronic signature). (In this context, we also mention the (however, already old) concept of steganography, where even the mere fact that a message has been transmitted (not only its contents) is to be kept secret. We will not discuss this subject further.) On the other hand, generalizations to multiparty systems also emerged. Nowadays, network security is a very important problem in practice.

A systematic introduction to the algebraic and number theoretic aspects was given in the Koblitz (1999) book "Algebraic Aspects of Cryptography". The goal of our text will be to give a similar insight into some probabilistic and statistical methods (in its broadest sense, so, for example, also using quantum stochastics) of cryptology. By no means do we claim completeness, only some introductions to certain topics can be given. Important areas, such as for example secret sharing, multi-party systems, zero-knowledge, problems on information transmission channels, linear cryptanalysis, digital fingerprinting, visual cryptography (see, for example, de Bonis, de Santis (2001)), etc., had to be (almost) entirely excluded. For further reading, we recommend that readers consult, in particular, the *Journal of Cryptology* and the various conference proceedings series, e.g., in the Springer Lecture Notes in Computer Science (EUROCRYPT, CRYPTO, ASIACRYPT, AUSCRYPT, INDOCRYPT, FAST SOFTWARE ENCRYPTION, etc.). What is also of interest are the journals *Designs, Codes, and Cryptography*, and *IEEE Transactions on Information Theory*, together with several "computational" periodicals. Sometimes, very important information can also be found in mathematical and stochastic journals/books, though this is rather the exception compared to the specific series devoted more to what is nowadays called "Theoretical Computer Science".

Book Structure

Let us now give a short description of the contents of the present book.

As already mentioned, in Section 1.1 we present the famous classical Vigenère system, which for a long time was believed to be as "secure as possible". Of course, no cryptosystem is absolutely secure in the literal sense of the word, since there is always the possibility of exhaustive search (in many cases, even though no better attack is known, however, also no *proof* that no better attack exists is available up to now). (Somewhat exceptional is quantum cryptography as it is briefly described in Chapter 13. But this is research in progress.) So actually the mere reasonable definition of "security" of a cryptosystem is a non-trivial task. In Section 1.2 we speak about the most natural (but expensive to realize) notion of "perfect secrecy", whereas other security concepts (weaker, but often more easily implementable and testable ones) are discussed in Sections 5.1 (Golomb's conditions, PN-sequences), 5.3 ("perfect pseudo-randomness", which means that a source cannot "efficiently" be distinguished from a truly random source), 5.4 (("almost") ideal local statistics), Chapter 10 ("semantic security", which is a "polynomially bounded" version of perfect secrecy in the sense that one assumes that the adversary has only "polynomial" computational resources), and Chapter 11 ("algorithmic complexity"). Of course, theoretically quite weak but in practice not unimportant is the requirement for maximal linear complexity (see Sections 5.1 and 7.11), if one confines oneself to linear feedback shift registers. A short remark follows about a misleading "intuitive" idea concerning cascade ciphers, against which Massey and Maurer (1993) warned in their paper "Cascade Ciphers: The Importance of Being First".

Chapter 2 is devoted to public-key ciphers, in particular to the RSA system. After the introduction of the RSA system, whose basis is the (probably true and therefore generally supposed) computational difficulty of factoring large integers, we present two of the best-known probabilistic primality tests (the Soloway-Strassen test, which, loosely speaking, tests Euler's criterion for the Legendre-Jacobi symbol, and the Rabin test, which is related to Fermat's

Little Theorem for residue rings modulo a prime). A specially designed probabilistic prime number test for numbers congruent 3 (mod.4) (i.e., candidates for prime factors of so-called Blum integers) has been presented by Müller (2003). In Section 2.4 we prove that in the RSA system, one has a "hard" least significant bit, which means that if ever one finds a probabilistic polynomial time algorithm for calculating the least significant bit of the plaintext from the public key and the ciphertext, then there exists also a probabilistic polynomial-time algorithm for reconstructing the whole plaintext from these data. "Hard bits" have been the subject of much subsequent literature. Another public-key algorithm, the Diffie-Hellman system, will be discussed in Chapter 8. Section 2.5 warns against careless hardware implementation, so that certain internal parameters (e.g., processing time) can be measured by the adversary, and advises on avoiding such attacks. For further reading about the subject of "timing attacks", we also refer to Schindler (2002a). In Section 2.6 we show how somebody can persuade his/her friend that he/she has found an RSA-secret key of somebody else without revealing any information about it, thus giving a first glimpse into the field of zero-knowledge proofs.

Chapter 3 presents Shor's algorithm (for whose invention Shor got the Nevanlinna prize) for factoring numbers with quantum computers. One must admit that up to now, quantum computers have been rather a theoretical concept and not yet producible in a usable way. The latest news about hardware research in this direction is rather pessimistic. Of course, from the viewpoint of users of classical cryptological devices this is reassuring, for if an adversary were really in possession of a quantum computer working on a large scale, then virtually all cryptosystems whose security is based on the "intractability" of the problem of factorizing numbers or the discrete logarithm problem would be breakable in "no" time (more precisely: in linear time, where up to now only behavior (e.g., for the quadratic or the number field sieve) of an order little better than exponential is known). We do not assume that the reader has any preliminary knowledge of quantum theory. All necessary explanations are given in Section 3.2. Shor's algorithm makes use of a result from the theory of continued fractions, which we will present in Section 3.3. Almost all cryptosystems work with keys, which, as a doctrine (at least in theoretical cryptology), is the only information on the cryptosystem that is assumed to (and can realistically) be kept secret. That is, one always assumes, in order to be on the safe side, that the adversary is in possession of the device that has been used for encryption/deciphering, but he has virtually no information about the key. The most secure way to provide a good key is to generate it with a genuine, physical generator, e.g., radioactive sources with Geiger counters or electronic noise produced by a semiconducting diode (see Chapter 4). For general use, for example, HOT BITS is a source of random bits stemming from beta radiation from the decay of krypton-85, and is available on the Internet. However, physical devices are very slow compared to pseudo-random generators, which we will treat in Chapter 5. Some considerations about possible constructions of good physical random number generators, such as some discussions on their quality due to Zeuner and the author, are the subject of Section 4.2. In Section 4.3 we address the general problem of obtaining random bits that are as unbiased as possible, if the disposable source only produces random bits with a certain bias. We will calculate the "extraction rate" (which indicates in some sense the asymptotical speed of the diminution of the bias per new random bit source, when the final output bit is produced by adding (mod.2) independent biased random bit sources) for rational biases. Interestingly enough, the extraction rate turns out to be independent of the size of the bias b, but to be determined solely by the arithmetic properties of b. However, one finds that the extraction rate is 0 for Lebesgue-almost all biases b.

On the contrary, we speak about pseudo-random generators in the following. In Chapter 5, we present some important examples (linear feedback shift registers (Section 5.1) and combinations thereof (Section 5.5), non-linear feedback shift registers (Section 5.4), shrinking and self-shrinking generators (Section 5.2), and the quadratic congruential generator (Section 5.6)).

Chapter 6 is a brief introduction to the most important notions of information theory as it is of use for us and to the aforementioned problem of authenticity. Section 6.3 is a new unorthodox approach.

In Chapter 7 we give a collection of some of the best-known tests for pseudo-random-number generators, orienting ourselves to a great extent at the tests suggested by Rukhin (2000a,b) and the test-battery used for evaluation of the AES. As is well-known, for a long time, the block cipher "data encryption standard" (DES) has been widely used, but, by using parallelism, it has been possible to break it. Then the NIST (National Institute of Standards and Technology) invited the worldwide cryptologic community to develop an "advanced encryption standard" (AES). The winner of this contest was the algorithm RIJNDAEL designed by Rijmen and Daemen.

Chapter 8 discusses the distribution of keys in the Diffie-Hellman public-key system. In this context, the notion of "strong primes" (primes p that are of the form $p = 2q + 1$ (where q is a prime)) is useful. Namely, it turns out that if the modulus is a strong prime, then the entropy of the Diffie-Hellman key is nearly the maximum possible, which means that it is recommendable to use strong primes as moduli. Similar considerations about bit security as we have in Section 2.4 apply for the Diffie-Hellman system, too. We refer to González Vasco, Shparlinski (2001).

Chapter 9 describes an attack on block ciphers that has become very popular in recent years, namely differential cryptanalysis. Roughly speaking, here the cryptanalyst makes use of cases where "differences/sums" (in the algebraic sense) of pairs of plaintexts leak through to differences/sums of the corresponding pairs of ciphertexts. In an iterative r-round block cipher, with this method it is sometimes possible to guess the r-th round subkey, then the

$(r-1)$-th round subkey, etc., iteratively until the whole key is found. Interestingly enough, although the theoretical results are generally proved under the assumption that the round keys are chosen as i.i.d. (independent and identically distributed), in practice they are experimentally verified (sometimes with even better behavior) if some key schedule algorithm is used. Section 9.2 generalizes distributional results for so-called characteristics (i.e., pairs of differences of plaintext/ciphertext pairs of bitstrings) due to Hawkes and O'Connor to residue rings of arbitrary modulus. Matsui (1994) developed the related concept of linear cryptanalysis, which we have excluded from our presentation.

In Chapter 10 we deal with semantic security. Roughly speaking, semantic security is a polynomially bounded variant of perfect security, i.e., one assumes that the adversary has only polynomially bounded resources.

A notion of "algorithmic complexity" (the so-called "Turing-Kolmogorov-Chaitin complexity", which is — roughly speaking — the length of the shortest program that one must feed to a universal Turing machine to generate as output a given bitstring) is considered in Chapter 11. However, this is of rather theoretical interest, since the algorithmic complexity of a given bitstring is not computable (in the sense of the Church Thesis). It turns out that in the sense of the Haar measure, for almost all bitstrings the algorithmic complexity is equal to the linear complexity, thus here we have a somewhat similar situation as for the extraction rate of biases in Section 4.3. At first glance this contradicts the fact that there are very simply constructed bitsequences with maximal linear complexity (e.g., 00...01), but the above-mentioned equivalence is not valid for "effectively constructible" sequences (see the title of the paper of Beth and Dai (1990): "If you can describe a sequence, it can't be random.").

Chapter 12 addresses the problem of collisions and the related "meet-in-the-middle" attack, which has to do with the well-known birthday paradox from probability theory.

Finally, we give a short glimpse into quantum cryptography in Chapter 13. In this situation, the receiver of an encrypted message will immediately detect (with arbitrarily large probability) if an adversary has manipulated the message (maybe even only "measured" it in the quantum-mechanical sense), which in general is of course not the case in classical cryptosystems. However, here also, the technology has not yet been developed far enough. Note that Chapter 13 deals with "genuine" quantum cryptography, whereas in Chapter 3 we showed how to solve a problem of classical cryptography by means of quantum computing.

Finally, a word about giving proper credits should be said: In cryptology, it is even more difficult than in other sciences to know to whom a certain result should really be attributed, since often methods that have been published later have already been developed (at least to a certain extent) before by cryptologists who were not allowed to publish their findings, especially during the time of the Second World War and the Cold War. So, citations of literature in our text should hardly be interpreted as a reference giving a credit to a certain person or group of persons. For example, one sees few Russian names occurring in the cryptological literature however, it turned out that Soviet cryptanalysts have had important successes in, for example, cryptanalysis, too.

In the body of this book, we give few formal citations, in order not to interrupt the smoothness of the presentation too much. Instead, we have included a section "Bibliographical Remarks" at the end of the text.

Chapters and sections with an asterisk treat more specific subjects and can be omitted at first reading.

About Notation and Terminology

Throughout the book, the symbol $I\!B$ will denote $GF(2) = \mathbb{Z}_2$, the field with the two elements 0 and 1, which will be called "bits" (exception: Section 4.3). Also, for a sequence $x = (x_1, x_2, \ldots)$, the symbol $x^{(n)}$ will mean the finite subsequence consisting of the first n elements: $x^{(n)} = (x_1, x_2, \ldots, x_n)$. The indicator function of the set B will be written as $\mathbf{1}(B)(.)$.

"W.l.o.g." means "without loss of generality". The shorthands "i.i.d." and "a.s." stand for the probabilistic notions "independent and identically distributed" and "almost surely" (i.e., "with probability one"). As already mentioned in the footnote at the beginning, the word "decipher" will mean the decoding of a ciphertext by its legitimate receiver, whereas "decrypt" is the breaking of the code by an adversary.

1 Classical Polyalphabetic Substitution Ciphers

1.1 The Vigenère Cipher

The classical situation in cryptology, which we will consider below, is the following: There are two parties, A (called "Alice" in the jargon) and B (called "Bob"). Alice would like to send a message to Bob by some channel. But this channel is unsecure because in-between the two, there is some adversary ("enemy", eavesdropper) E (called "Eve") who either wants

- to listen in on the message sent from A to B and/or
- to send a message herself to B, asserting that this message comes from A and/or
- to change a message indeed sent by A to B.

All these three attacks should be avoided. The first attack (listening in) concerns the problem of secrecy (or confidentiality), the second that of authenticity, and the third that of integrity. In other words, there are two independent goals: To reach secrecy resp. authenticity/integrity, the output resp. input of the channel from A to B should be exclusive. Of course, there are more general cryptologic situations (multi-party models, secret sharing, zero-knowledge, etc.). But these will not be considered here (except in the short Section 2.6). Also the integrity/authenticity problem will only be addressed in Sections 2.1 (electronic RSA signature) and 6.2 (impersonation attack), and Chapter 12 (meet-in-the-middle attack). Apart from that, in this introductory text we will mainly be concerned with secret keeping.

In this chapter, we will present a classical cryptosystem, the so-called Vigenère cipher, invented in 1586 by the French diplomat Blaise de Vigenère (1523-1596). It belongs to the class of polyalphabetic cryptosystems, which means that the same letter of plaintext is not always encoded by the same letter of ciphertext. This fact is of great importance in general. If a cryptosystem is monoalphabetic, i.e. if every letter of plaintext is always encrypted by the same letter of ciphertext, then statistical properties of the letters of the language in which the plaintext is written automatically leak through to the ciphertext, i.e. (for long enough messages) frequent letters (or m-grams) in the ciphertext correspond to frequent letters (or m-grams) in the plaintext, and by some statistical analysis it is, in general, not too difficult to find the

D. Neuenschwander: Prob. and Stat. Methods in Cryptology, LNCS 3028, pp. 9-15, 2004.
© Springer-Verlag Berlin Heidelberg 2004

plain-/ciphertext correspondence of frequent letters (m-grams) of the language. To fill in the rest, often some "trial and error" helps (in particular with some additional information about "mots probables" (words that are likely to occur in the message)).

The Vigenère system is very simple and works as follows: Given a keyword, e.g., "PEACE" and the plaintext

OSAMABINLADEN,

then one writes the plaintext and the repeated keyword under each other and "adds" the corresponding letters mod.26 (where A is interpreted as 0, B as 1, etc.) to obtain the ciphertext:

Plaintext	O	S	A	M	A	B	I	N	L	A	D	E	N
Keyword	P	E	A	C	E	P	E	A	C	E	P	E	A
Ciphertext	D	W	A	O	E	Q	M	N	N	E	S	I	N

If Bob knows the key word, he can retrieve the plaintext from the ciphertext simply by subtracting the corresponding letters of the keyword mod. 26. But what cryptanalysis is concerned, one must say that although this system is polyalphabetic as such, always after k places (if k is the length of the keyword) the same substituting alphabet (which is even just a shift of the original alphabet in the sense of its interpretation as elements of $\mathbb{Z}_{26}$) is used. This gives rise to an algebraic method (the so-called Kasiski test) of determining the keyword length up to multiples. Together with the stochastic Friedman test, which yields the order of magnitude of the length of the keyword, one can determine in most cases the actual length of the keyword. If this is known, for every place modulo the length of the keyword, one must replace the letter of the ciphertext that occurs most frequently by some very frequent letter of the language in which the plaintext is written to determine the shift, and then with little routine work one can then (in general) reconstruct the plaintext thus. Let us describe the details: The Kasiski test is named after the Prussian major Friedrich Wilhelm Kasiski (1805-1881), although it had been found nine years before him (but had not been published) by Charles Babbage (1792-1871) in 1854. It rests on the following observation: If a certain word (for example a preposition or a conjunction, etc.) occurs several times in the plaintext and if by chance (which is often quite large) the distance between two such occurrences of the same word is a multiple of the length of the keyword, then this word is encoded both times by the same sequence of letters in the ciphertext. Or - spoken the other way round - if one detects the same subsequences of letters (maybe even short ones, e.g., of length 3) several times in the ciphertext, then the distance between them is quite probably a multiple of the keyword length. Now the second part will be a little more involved, it is the so-called Friedman test, which was developed by William Friedman in 1925. This is a test zhat is of stochastic nature. Consider a plaintext of n letters, built from the Latin alphabet with the 26 characters "A", "B",.... Let n_1 be the number of "A"s, n_2 the number of "B"s, etc. in the plaintext (hence $n = \sum_{i=1}^{26} n_i$). Then the index of coincidence I is defined as the probability that an arbitrary pair of letters taken from the plaintext consists of the same 2 letters, i.e.

$$I = \frac{\sum_{i=1}^{26} n_i(n_i - 1)}{n(n-1)}.$$

If p_i denotes the probability that on some fixed place (in a text of the considered language) letter i occurs, then (if the text is long enough) we have

$$I \approx \sum_{i=1}^{26} p_i^2. \tag{1.1}$$

The expression on the right-hand side of (1.1) decreases, if the distribution of the letters in the language becomes more regular and takes its minimum 0.0385 if $p_i = 1/26$ for all $i \in \{1, 2, \ldots, 26\}$. The index of coincidence of a natural language typically has about the double value (e.g. about 0.0667 for English). With a monoalphabetic substitution, the index of coincidence remains unchanged whereas it decreases (in general) with a polyalphabetic substitution. So a coincidence index of a polyalphabetic substitution tends to be low (near 0.0385), whereas a significantly higher value suggests that a monoalphabetic substitution method has been used. Now I (from the ciphertext) can be used to determine the approximate length of the keyword as follows: Assume the keyword has length ℓ (and, for simplicity, that n is w.l.o.g. a multiple of ℓ). Then write a $((n/\ell) \times \ell)$-matrix M where the letters number $k + j\ell$ ($j = 0, 1, 2, \ldots, (n/\ell) - 1$) of the ciphertext form the k-th column. Now if we take a (random) pair of letters in some fixed column, the probability that both letters are equal is about (in practice a little more than) 0.0667, since the individual columns have been encrypted monoalphabetically. The number of pairs of two letters of the same column is given by $n((n/\ell) - 1)/2$. If we take random pairs of letters of two different columns, the probability of obtaining the same letter twice is about 0.0385 (if the keyword is "long" and "random" enough). The number of pairs from two different columns is $n(n - (n/\ell))/(2\ell)$. Hence the probability p to have equal letters if one takes a pair of two letters from the matrix M at random is about

$$p = \frac{\frac{n(n-\ell)}{2\ell} \cdot 0.0667 + \frac{n^2(\ell-1)}{2\ell} \cdot 0.0385}{n(n-1)/2}$$

$$= \frac{1}{\ell/(n-1)}(0.0282n + \ell(0.0385n - 0.0667)).$$

Since this expression is an approximation for I from the ciphertext, we may replace p by I from the ciphertext and by solving with respect to ℓ we obtain Friedman's formula for the approximate keyword length ℓ:

$$\ell = \frac{0.0282n}{(n-1)I - 0.0385n + 0.0667}, \tag{1.2}$$

where I is the empirical coincidence index of the ciphertext.

1.2 The One Time Pad, Perfect Secrecy, and Cascade Ciphers

The method of attack described in the foregoing section becomes more and more difficult if the keyword becomes longer and longer and is "random enough". If, as a keyword, one takes a random string of the same length as the plaintext itself, then the ciphertext becomes a random string, too, and thus the system is theoretically (or "perfectly") secret (or "secure"). This system is called the One-Time Pad and was invented in 1917 by G. S. Vernam (1890-1960) (that is why it is also called the "Vernam cipher"). But what is the practicability of it, if the key (which has also to be transferred once from Alice to Bob) must have the same length as the plaintext? Do we really gain something? The anwer is yes, for the key can be exchanged at any time *before* the transmission of the message becomes necessary, e.g. by some trustworthy courier. But it is important that any key is used only once (and then destroyed), for if two messages $x_1 x_2 \ldots x_n$ and $x_1' x_2' \ldots x_n'$ have been encrypted by the key $z_1 z_2 \ldots z_n$ to give the ciphertexts $y_1 y_2 \ldots y_n$, resp. $y_1' y_2' \ldots y_n'$, then $y_i + y_i' = x_i + x_i'$. So immediately the sum of the two plaintexts is already known, which reveals a lot of information!

Let us discuss the notion of perfect secrecy in some more detail.

Definition 1.1. *A cryptosystem is said to have perfect secrecy if for all plaintexts X and all ciphertexts Y, we have*

$$P(X|Y) = P(X).$$

Generally, perfectly secret cryptosystems can be characterized as follows:

Theorem 1.1. *Assume $P(X) > 0$ for any plaintext X and assume that the key space has the same size as the space of possible ciphertexts. Then a cryptosystem has perfect secrecy iff the distribution over the key space is uniform and if for any plaintext X and any ciphertext Y there is exactly one key Z that encrypts X to Y.*

Proof: 1. We first prove the "only if"-direction. Let X denote a plaintext and assume there is a ciphertext Y such that there is no key Z that encrypts X to Y. Then

$$P(X|Y) = 0 < P(X),$$

which contradicts the definition of perfect secrecy, so at least one key Z encrypting X to Y must exist. But since by the assumption there are exactly as many keys as ciphertexts, Z must be unique. It remains to prove the uniformity of the distribution of the keys. Denote by $Z(X)$ the key that encrypts the plaintext X to the ciphertext Y. By Bayes' rule, we have

$$P(X|Y) = \frac{P(Y|X)P(X)}{P(Y)} = \frac{P(Z(X))P(X)}{P(Y)}. \tag{1.3}$$

By perfect secrecy, $P(X|Y) = P(X)$, so that (1.3) implies $P(Z(X)) = P(Y)$. So $P(Z(X))$ is the same for any plaintext X, and uniformity follows from the fact that any key Z has the property $Z = Z(X)$ for some plaintext X.

2. Now we pass to the "if"-part. For all X, Y there is exactly one key $Z = Z(X, Y)$ that encrypts X to Y. Again by Bayes'rule (as in (1.3))

$$P(X|Y) = \frac{P(X)P(Y|X)}{P(Y)}$$
$$= \frac{P(X)P(Z(X,Y))}{\sum_{X'} P(X')P(Z(X',Y))} \tag{1.4}$$

(where the sum in the denominator runs over all plaintexts X') and the fact that all $P(Z(X,Y))$ are equal, we obtain that the denominator in (1.4) is equal to the reciprocal value of the size of the key space and hence $P(X|Y) = P(X)$. $\square$

A notion related to perfect secrecy is semantic security, which will be treated in more detail in Chapter 10. The effect of perfect secrecy is that the adversary, even if he has unlimited computer resources, can gain no information about the plaintext from the ciphertext, except its length if this is not a known parameter (see Theorem 10.1). The disadvantage of the requirement of perfect secrecy is that the key must be at least as long as the plaintext. Roughly speaking, semantic security is a polynomially bounded variant of perfect secrecy, i.e. one assumes that the adversary has only polynomially bounded computer resources.

A word about cascade ciphers: A cascade cipher is a sequence of component ciphers C_i ($i = 1, 2, \ldots, r$), where the output of Y_i of cipher C_i is used as input X_{i+1} for cipher C_{i+1}. In every component cipher, a key Z_i is used:

$$Y_i = C_i(X_i, Z_i) = C_i(Y_{i-1}, Z_i)$$

It is assumed that the keys $Z_1, Z_2, \ldots, Z_r$ are statistically independent (otherwise one speaks of a product cipher). So the input X for the whole cascade cipher is $X = X_1$, whereas the output is $Y = Y_r$. Now one is tempted to believe that a cascade cipher is at least as hard to break as its hardest component. But as Massey and Maurer (1993) have shown, this is only true for "pure" known-plaintext, chosen-plaintext, and chosen-ciphertext attacks in which Eve can not make use of information about the statistics of the plaintext. As soon as the statistics of the plaintext is known, a cascade cipher can possibly be easier to break than its hardest component, as the following counterexample shows: Let C_1, C_2 be two block ciphers with input/output alphabet consisting of the 4 letters A,B,C,D. Assume that the keys Z_1 and Z_2 are independent unbiased random bits. The component ciphers C_i transform the alphabet as follows (by a little free use of notation):

$$C_1((A, B, C, D), 0) := (C, D, A, B),$$

$$C_1((A, B, C, D), 1) := (C, D, B, A),$$

$$C_2((A, B, C, D), 0) := (C, D, A, B),$$

$$C_2((A, B, C, D), 1) := (D, C, A, B).$$

Now we assume that for the plaintext statistics we have $P(C) = P(D) = 0$. Then C_1 is completely insecure for this plaintext source, but C_2 is perfectly secret since the plaintext and the ciphertext are statistically independent. But on the other hand, the cascade cipher $C_2 \circ C_1$ is completely insecure, since it is just the identity transformation on $\{A, B\}$! What one can only prove is that a cascade cipher is at least as secure as the *first* component cipher (see Massey, Maurer (1993). "Cascade ciphers: The importance of being first"). If $C_1 = C_2 = \ldots = C_r$, then of course (since the components commute), the iteration cipher is at least as secure as the component ciphers themselves. This setup will be considered in more detail in Chapter 9.

Theorem 1.2. *A cascade of n ciphers is at least as difficult to break as the first component.*

Proof: Consider an oracle that gives, upon request, the keys of all component ciphers in the cascade except the key of the first component. Breaking the cascade with the oracle's help can not be more difficult than breaking it without this help because the oracle's information can always be disregarded. However, breaking the cascade with the oracle's help is equivalent to breaking the first component cipher with the oracle's help because on the one hand every cryptogram of the cascade can with assumed negligible computation be converted into the corresponding cryptogram for the first component cipher and vice versa, and on the other hand the plaintexts of the first component cipher and the cascade are the same. However, since the information provided by the oracle is statistically independent of the first key, it follows that breaking only the first component cipher with the oracle's help is equivalent to breaking this first component without the oracle's help. Or - in other words - it follows from the fact that if the cryptanalyst (Eve) attacking the first component cipher wishes to embed that component cipher in an artificial cascade in which she herself chooses the second and all subsequent keys (independently of the first key by assumption) so as to avail herself of the oracle's aid, then she already possesses all the information that the oracle can provide. So breaking the first component cipher can not be more difficult than breaking the whole cascade cipher. $\square$.

2 RSA and Probabilistic Prime Number Tests

2.1 General Considerations and the RSA System

The RSA cryptosystem (named after R. Rivest, A. Shamir, and L. Adleman, who published it in the 1970s) is one of the best-known so-called public key cryptosystems. The idea is the following: Every participant chooses two different big primes p and q "at random" and calculates their product $n = pq$. Then he chooses some arbitrary natural number e that is relatively prime to the Euler totient function $\varphi(n)$ (which denotes the number of relative primes to n that are smaller than n or - in other words - the number of invertible elements mod.n). In our situation, we have $\varphi(n) = (p-1)(q-1)$. So for e one can take, e.g., any prime larger than $(p-1)(q-1)$ or, what makes the decoding and encryption in the binary system especially simple, the 4th Fermat Number $F_4 := 2^{2^4} + 1 = 65'537 \ (= 1'0000'0000'0000'0001$ in the binary system). The pair (n, e) is the so-called public key of the participant, which he publishes and will be known to everybody. As his secret key, he keeps the solution $d < \varphi(n)$ of the equation

$$ed = 1(mod.(p-1)(q-1)). \tag{2.1}$$

This solution can be found rapidly by the Euclidean algorithm if p and q are known. But factorizing numbers n seems to be computationally hard in the sense that there seems to exist no algorithm that is faster than exhaustive search. Moreover, there is no known algorithm to solve (2.1) faster than by finding p and q. But the actual equivalence has not been proved up to now. See also Boneh, Venkatesan (1998). There are similar systems (however, with other disadvantages) where breaking the system is provably equivalent to finding the secret key, for example the Rabin system (Kranakis (1986)) or the Williams (1980) algorithm. For convenience, we will now write (n_A, e_A) and (n_B, e_B) for the public key of Alice and Bob resp., and d_A and d_B for their respective secret keys. Assume Alice wants to send a message x (w.l.o.g. in the form of a natural number mod.n_B) to Bob. For that, she calculates the ciphertext y (which will also be a natural number mod.n_B) by

$$y := x^{e_B}(mod.n_B) \tag{2.2}$$

D. Neuenschwander: Prob. and Stat. Methods in Cryptology, LNCS 3028, pp. 17-35, 2004.
© Springer-Verlag Berlin Heidelberg 2004

and sends this to Bob. Bob will make the decoding

$$x = y^{d_B}(mod.n_B) \qquad (2.3)$$

(which follows from (2.2) by (2.1) and Fermat's Little Theorem). So the RSA system seems to ensure confidentiality. The system can also be used to ensure authenticity: For that, Alice sends, in addition to the encrypted message x, her "electronic signature" m, encrypted by

$$u := m^{d_A}(mod.n_A), \qquad (2.4)$$

to Bob. Finding d_A from u is the so-called discrete logarithm problem, which is also believed to be hard. So by signing, Alice does not reveal her private key d_A. Since d_A is only known to Alice, she alone can have produced u, so u has really the role of a "signature". On the other hand, Bob can verify that this is really Alice's signature by checking if

$$u^{e_A} \stackrel{?}{=} m(mod.n_A). \qquad (2.5)$$

A probabilistic (or so-called Monte Carlo) primality test is an algorithm $A_P(n)$ that, for the input n, gives one of the two answers "prime" or "composite" such that if it yields "composite", then n is composite and if it yields "prime", then n is indeed prime with high probability. It seems to be a general fact in prime number testing that if in the case of the output "prime" one is satisfied that this answer is correct only up to some small error probability, then the test runs much faster, or - in other words - what costs most effort is to obtain absolute security in improbable cases. At least theoretically, a major breakthrough has been achieved recently by Agrawal et al. (2003), who gave an unconditional (i.e., not depending on any unproven assumption as, e.g., the Extended Riemann Hypothesis (see Section 2.2) deterministic polynomial-time algorithm to decide whether or not a number is prime (see also Bornemann (2002), Bernstein (2002), and New York Times 8/8/2002).

In detail, for a probabilistic primality test one defines a so-called primality sequence $P = \{P_n\}_{n\geq 1}$ of sets of natural numbers with the following properties:

(i) $P_n \subset \mathbb{Z}_n^*$ (= group of integers mod.n relatively prime to n).

(ii)   Given $b \in \mathbb{Z}_n^*$ one may check in time polynomial in the length of the binary expansion of n if $b \in P_n$.

(iii)   If n is prime, then $P_n = \emptyset$.

(Iv)   There exists a so-called primality constant $\varepsilon \in]0,1[$ (independent of n) such that for all sufficiently large composite odd $n \geq 1$ one has $P(x \in \mathbb{Z}_n^* : x \notin P_n) \leq \varepsilon$.

Now the test algorithm works as follows:

– Input: $n \geq 2$.

– Choose an integer $x \in \mathbb{Z}_n^*$ at random.

– Output: $A_P(n) =$"prime" if $x \notin P_n$ and $A_P(n) =$"composite" if $x \in P_n$.

If the test is run sufficiently many times (with independent values for x), then the error probability can be made arbitrarily small:

$$P(A_P(n) = \text{"prime"}, \text{ although } n \text{ is composite}) \leq \varepsilon^m.$$

2.2 The Solovay-Strassen Test

This test uses a well-known object from number theory, the so-called Legendre-Jacobi symbol $(x|n)$. If p is a prime and $x \in \mathbb{Z}_p^*$, then the Legendre symbol is defined as $(x|p) = 1$ if x is a quadratic residue modulo p and $(x|p) = -1$ else. By Euler's criterion (see, e.g., Kranakis (1986), Theorem 1.11), for all odd primes p one can calculate the Legendre symbol explicitly as

$$(x|p) = x^{(p-1)/2}(mod.p).$$

Now, for general n and $x \in \mathbb{Z}_n^*$, one defines the Legendre-Jacobi symbol by

$$(x|n) = \prod_{i=1}^{t}(x|p_i),$$

if $n = \prod_{i=1}^{t} p_i$ denotes the prime factorization of n.

Now the primality sequence of the Solovay-Strassen test is defined based on Euler's criterion:

$$P_n = \{x \in \mathbb{Z}_n^* : x^{(n-1)/2} \neq (x|n)(mod.n)\}.$$

From Euler's criterion, conditions (i)-(iii) for primality sequences are fulfilled. It remains to prove (iv). For this, we need some preparation.
Denote by $\nu_m(t)$ the largest k such that $m^k | t$.

Lemma 2.1. Let $n = \prod_{1=1}^{t} p_i^{k_i}$ be the prime factorization of the odd integer n (i.e., the p_i are the different prime factors of n) and $m \in \mathbb{N}$. Put $\nu := \min\{\nu_2(p_i - 1) : i = 1, 2, \ldots, t\}$ and $s := \prod_{i=1}^{t} \gcd(m, \varphi(p_i^{k_i}))$. Then
(1)   The equation $x^m = 1(mod.n)$ has exactly s solutions.
(2)   There exists some x with $x^m = -1(mod.n)$ iff $\nu_2(m) < \min\{\nu_2(p_i-1) : i = 1, 2, \ldots, t\}$
(3)   If the equation $x^m = -1(mod.n)$ has a solution, then it has exactly s solutions.

Proof: Let (for a prime p and a generator g of $\mathbb{Z}_p^*$) the shorthand $\text{index}_{p,g}(x)$ denote the unique $m \leq p - 2$ such that $x = g^m(mod.p)$. For each $i \in \{1, 2, \ldots, t\}$, let g_i be a generator of $\mathbb{Z}_{p_i^{k_i}}^*$. Taking indexes on both sides of the equation $x^m = a(mod.n)$ one gets

$$m \cdot \text{index}_{p_i, g_i}(x) = \text{index}_{p_i, g_i}(a)(mod.\varphi(p_i^{k_i})).$$

Substituting $a = 1$ yields

$$m \cdot \text{index}_{p_i, g_i}(x) = 0 (mod.\varphi(p_i^{k_i})), \qquad (2.6)$$

whereas for $a = -1$ we get $\text{index}_{p_i, g_i}(-1) = \varphi(p_i^{k_i})/2$ and thus

$$m \cdot \text{index}_{p_i, g_i}(x) = \varphi(p_i^{k_i})/2 (mod.p_i^{k_i}). \qquad (2.7)$$

Now (1) of Lemma 2.1 follows from (2.6) and the theorem on the solution of linear congruences. The same theorem also implies that (2.7) has a solution iff

$$\gcd(m, \varphi(p_i^{k_i})) | \varphi(p_i^{k_i})/2$$

for all $i = 1, 2, \ldots, t$. But the latter holds exactly iff $\nu_2(m) < \min\{\nu_2(p_i - 1) : i = 1, 2, \ldots, t\}$. $\square$

The next is a lemma due to Monier:

Lemma 2.2. *Let n be odd and assume $p_1, p_2, \ldots, p_t$ are the distinct prime factors of n. Then one can write*

$$|\mathbb{Z}_n^* \backslash P_n| = \delta_n \prod_{i=1}^{t} \gcd(\frac{n-1}{2}, p_i - 1)$$

where δ_n assumes one of the values $1/2, 1, 2$.

Proof: Define the multiplicative group endomorphisms f_n, g_n, h_n of $\mathbb{Z}_n^*$

$$f_n(x) := x^{(n-1)/2}(mod.n),$$

$$g_n(x) := (x|n)(mod.n),$$

$$h_n(x) := (x|n) \cdot x^{(n-1)/2}(mod.n).$$

Let K_n, L_n, M_n be the kernels of f_n, g_n, h_n, resp., and denote

$$K_n' := \{x \in \mathbb{Z}_n^* : f_n(x) = -1(mod.n)\},$$

$$L_n' := \{x \in \mathbb{Z}_n^* : g_n(x) = -1(mod.n)\},$$

$$M_n' := \{x \in \mathbb{Z}_n^* : h_n(x) = -1(mod.n)\}.$$

Clearly $M_n = \mathbb{Z}_n^* \backslash P_n$. By Lemma 2.1 it follows that

$$|K_n| = \prod_{i=1}^{t} \gcd(\frac{n-1}{2}, p_i - 1).$$

However, $M_n = (K_n \cap L_n) \cup (K_n' \cap L_n')$. Thus

$$|M_n| = \begin{cases} |K_n \cap L_n| & : & K_n' \cap L_n' = \emptyset \\ 2|K_n \cap L_n| & : & K_n' \cap L_n' \neq \emptyset \end{cases}$$

(if $K'_n \cap L'_n \neq \emptyset$, then choose $x_0 \in K'_n \cap L'_n$ and consider the bijection $x \to xx_0$ to prove $|K_n \cap L_n| = |K'_n \cap L'_n|$). A similar argument using the decomposition $K_n = (K_n \cap L_n) \cup (K_n \cap L'_n)$ can be used to show

$$|K_n \cap L_n| = \left\{ \begin{array}{ll} |K_n| & : \quad K_n \cap L'_n = \emptyset \\ (1/2)|K_n| & : \quad K_n \cap L'_n \neq \emptyset. \end{array} \right.$$

The assertion follows. $\square$

Theorem 2.1. *For all composite odd integers n we have*

$$\frac{|\mathbb{Z}_n^* \backslash P_n|}{\varphi(n)} \leq \frac{1}{2}.$$

Proof: Let again $\prod_{i=1}^{t} p_i^{k_i}$ be the prime factorization of n (i.e., $p_1, p_2, \ldots, p_t$ the distinct prime factors of n). By Lemma 2.2 it follows that

$$\frac{|\mathbb{Z}_n^* \backslash P_n|}{\varphi(n)} \leq \delta_n \prod_{i=1}^{t} \frac{\gcd(\frac{n-1}{2}, p_i - 1)}{p_i^{k_i-1}(p_i - 1)}. \tag{2.8}$$

If for some i it holds that $k_i \geq 2$, then the right-hand side of (2.8) is bounded from above by $\delta_n/3 \leq 2/3$. So $\mathbb{Z}_n^* \backslash P_n$ must be a proper subgroup of $\mathbb{Z}_n^*$ and hence $|\mathbb{Z}_n^* \backslash P_n| \leq (1/2)\varphi(n)$. Thus w.l.o.g. we may assume that all $k_i = 1$. Assume $\mathbb{Z}_n^* = M_n$. Since n is composite, it follows that $t \geq 2$. Assume g is a generator of $\mathbb{Z}_{p_1}^*$. By the Chinese remainder theorem there exists an $x \in \mathbb{Z}_n^*$ with $x = g(mod.p_1)$ and $x = 1(mod.(n/p_1))$. By the assumption $\mathbb{Z}_n^* = M_n$ it follows that $x^{(n-1)/2} = (x|n)(mod.n)$. But $(x|n) = \prod_{i=1}^{t}(x|p_i) = (g|p_1) = -1$. So $x^{(n-1)/2} = -1(mod.(n/p_1))$, which is a contradiction to $x = 1(mod.(n/p_1))$.$\square$

We mention that the Solovay-Strassen test is deterministic if the so-called Extended Riemann Hypothesis (see, e.g., Kranakis (1986), 2.10), a famous conjecture in analytic number theory, is true. This conjecture asserts the following: Let χ be a so-called character modulo n, i.e., a group homomorphism $\chi : \mathbb{Z}_n^* \to \mathbb{C}^*$, extended to $\mathbb{N}$ by $\chi(x) := 0$ if $\gcd(x, n) \neq 1$. Then the Dirichlet L-series with respect to the character χ is defined as

$$L_\chi(z) := \sum_{k=1}^{\infty} \frac{\chi(k)}{k^z},$$

which is convergent for all complex z with real part greater than 1 and can be meromorphically extended to an analytic function for all complex z with positive real part. Now the Extended Riemann Hypothesis is the conjecture that all zeroes of L_χ with real part in $]0, 1]$ have in fact real part $1/2$. Up to now, the Extended Riemann Hypothesis has not yet been proved, but there is overwhelming evidence (both by theoretical arguments and numerical calculations) that it really holds (see e.g. Odlyzko (2001)).

2.3 Rabin's Test

Here, the primality sequence is defined as

$$P_n = \{x \in \mathbb{Z}_n^* : x^{(n-1)/2^e} \neq 1 (mod.n) \text{ and } x^{(n-1)/2^h} \neq -1 (mod.n)$$
$$\text{for all } 0 < h \leq e\},$$

where $e := \nu_2(n-1)$ (as defined before). Also here, properties (i)-(iii) of primality sequences are easily verified. We must again prove (iv). We have

$$\mathbb{Z}_n^* \backslash P_n = \{x \in \mathbb{Z}_n^* : x^{(n-1)/2^e} = 1 (mod.n) \text{ or } x^{(n-1)/2^h}$$
$$= -1 (mod.n) \text{ for some } 0 < h \leq e\}. \tag{2.9}$$

Again, a lemma due to Monier determines the exact size of this set:

Lemma 2.3. *Assume n is a composite odd integer with prime factorization $n = \prod_{i=1}^t p_i^{k_i}$ (p_i distinct primes). If we write $n - 1 = 2^e u$ (u odd), $p_i - 1 = 2^{\mu_i} u_i$ (u_i odd), and $\mu := \min\{\mu_1, \mu_2, \ldots, \mu_t\}$, then*

$$|\mathbb{Z}_n^* \backslash P_n| = (1 + \frac{2^{t\mu} - 1}{2^t - 1}) \prod_{i=1}^t \gcd(u, u_i).$$

Proof: Put $s := \prod_{i=1}^t \gcd(u, u_i)$. By Lemma 2.1, the first congruence in (2.9) has exactly s solutions. For any given h, the other congruence in (2.9) has a solution (and thus s solutions) iff $\nu_2((n-1)/2^h) = e - h < \mu$. So, for each $h > e - \mu$, the number of solutions of the equation

$$x^{(n-1)/2^h} = -1 (mod.n)$$

is given by

$$\prod_{i=1}^t \gcd(\frac{n-1}{2^h}, p_i - 1).$$

Hence

$$|\mathbb{Z}_n^* \backslash P_n| = s + \sum_{h=e-\mu+1}^e \prod_{i=1}^t \gcd(\frac{n-1}{2^h}, p_i - 1).$$

Now the assertion follows from the fact that

$$\gcd(\frac{n-1}{2^h}, p_i - 1) = 2^{e-h} \gcd(u, u_i). \square$$

Let e and n be as in Lemma 2.3 and define the set

$$R_n := \{x \in \mathbb{Z}_n^* : x^{n-1} \neq 1 (mod.n) \text{ or } 1 < \gcd(x^{(n-1)2^h} - 1, n) < n$$
$$\text{for some } 0 \leq h \leq e\}.$$

The following lemma is due to Miller, Rabin, and Monier:

Lemma 2.4. *For all odd integers $n > 2$, we have $P_n = R_n$.*

Proof: Take an arbitrary $x \in \mathbb{Z}_n^*$ and consider, for each h such that $2^h | n - 1$, the expressions

$$d(h) := \frac{n-1}{2^h},$$

$$b(h) := x^{d(h)},$$

and

$$g(h) := \gcd(b(h) - 1, n).$$

Then, if $2^h | n - 1$, we have the properties:

$$g(h) = n \iff b(h) = 1 (mod.n), \tag{2.10}$$

$$g(h) = n \implies g(h - 1) = n, \tag{2.11}$$

and

$$b(h - 1) = b(h)^2. \tag{2.12}$$

1. We first prove that $P_n \subset R_n$. Assume the contrary and let $x \in P_n \backslash R_n$. It follows that there must be an integer $k \leq e$ with $g(k) = n$. As $x \in P_n$, we have $x^{d(e)} \neq 1 (mod.n)$, so $g(e) \neq n$. Hence, there exists a $k < e$ with the property

$$g(0) = g(1) = \ldots = g(k) = n > g(k + 1) = g(k + 2) = \ldots = g(e) = 1.$$

Hence $b(k + 1)^2 = 1 (mod.n)$ and thus $n | (b(k + 1) - 1)(b(k + 1) + 1)$. Together with the fact that $g(k + 1) = \gcd(b(k + 1) - 1, n) = 1$ this yields that $b(k + 1) = -1 \quad (mod.n)$, which contradicts the assumption $x \in P_n$.

2. Now let us show the relation $R_n \subset P_n$. Assume, on the contrary, that there exists some $x \in R_n \backslash P_n$. Then either $b(e) = 1 (mod.n)$ or there is some $h \in \{1, 2, \ldots, e\}$ with $b(h) = -1 (mod.n)$. In the first case $x \notin R_n$, so we may assume that $b(e) \neq 1 (mod.n)$. We may choose some $k \leq e$ such that

$$b(0) = b(1) = \ldots = b(k - 1) = 1 (mod.n),$$

but

$$b(k) = -1 (mod.n).$$

From the fact that $b(k) = b(k + j)^{2^j} = -1 (mod.n)$, we get

$$b(k) - 1 = b(k + 1)^2 - 1 = b(k + 2)^4 - 1 = \ldots = b(e)^{2^{e-k}} - 1 = -2 (mod.n).$$

But for all $j \leq e - k$ there is an integer s_j with the property

$$b(k + j)^{2^{2^j}} - 1 = (b(k + j) - 1)s_j = -2 (mod.n).$$

As n is odd and greater than 2 by assumption, we obtain that $g(k+j) = 1$ for all $j \le e - k$. Since, on the other hand, $g(j) = n$ for all $j < k$, we deduce that $g(h) \in \{1, n\}$ for all h. So indeed $x \notin R_n$. □

Now we are ready to calculate the primality constant, i.e., to verify property (iv) of primality sequences. Denote $q_i := p_i^{k_i}$ (as before, p_i denote the distinct prime factors of n and k_i the maximal power in which they occur in the prime factorization of n ($i = 1, 2, \ldots, t$)). Furthermore, put $h_i := \gcd(\varphi(q_i), n - 1)$, $m_i := \varphi(q_i)/h_i$, $e_i := \nu_2(h_i)$, and $\alpha_i := \max\{e_i - e_j : j = 1, 2, \ldots, t\}$. One observes that if $e_i = \min\{e_1, e_2, \ldots, e_t\}$, then $\alpha_i = 0$. Define

$$I := \{1 \le i \le t : \alpha_i > 0\},$$

$$J := \{1 \le i \le t : \alpha_i = 0\},$$

and $\alpha := \sum_{i=1}^{t} \alpha_i$, $\beta := |J|$. We have $\beta > 0$ and $\alpha + \beta \ge t$. The following theorem gives a general expression for the primality constant of the Rabin test:

Theorem 2.2. *If $n > 2$ is a composite odd integer with prime factorization $n = \prod_{i=1}^{t} p_i^{k_i}$ (p_i the distinct prime factors), then*

$$\frac{|\mathbb{Z}_n^* \backslash R_n|}{\varphi(n)} \le \frac{1}{2^{\alpha+\beta-1} \prod_{i=1}^{t} m_i}.$$

Proof: Assume $x \in \mathbb{Z}_n^* \backslash R_n$. So $x^{n-1} = 1 (mod.n)$. For $i = 1, 2, \ldots, t$, denote by g_i a generator of $\mathbb{Z}_{q_i}^*$. It follows that there are $s_i < \varphi(q_i)$ with $x = g_i^{s_i} (mod.q_i)$. Hence $x^{n-1} = g_i^{s_i(n-1)} = 1 (mod.q_i)$ and $\varphi(q_i)|s_i(n-1)$. As $\gcd(m_i, n-1) = 1$ and $m_i|s_i(n-1)$, we obtain the existence of $\ell_i < \varphi(q_i)/m_i$ such that $s_i = m_i \ell_i$. Hence

$$x = g_i^{m_i \ell_i} (mod.q_i) \tag{2.13}$$

and $s_i(n-1) = m_i \ell_i(n-1) = \varphi(q_i)\ell_i \frac{n-1}{h_i}$. It can be proved that

$$2^{\alpha_i}|\ell_i \quad (i = 1, 2, \ldots, t). \tag{2.14}$$

[W.l.o.g. we may assume $\alpha_i > 0$. Choose j such that $\alpha_i = e_i - e_j > 0$ and define $f_i := \nu_2(n-1) - e_i \ge 0$. Thus for $\gamma_i := e_i - e_j + f_i$ we get $\nu_2(d(\gamma_i)) = e_j$. Furthermore, $h_j|d(\gamma_i)$. Thus $\varphi(q_j) = h_j m_j | m_j d(\gamma_i)$. From (2.13) we get

$$x^{d(\gamma_i)} = g_j^{m_j \ell_j d(\gamma_i)} = 1 (mod.q_j).$$

Hence $1 < q_j \le \gcd(x^{d(\gamma_i)} - 1, n) = n$ (since $x \notin R_n$) and thus $x^{d(\gamma_i)} = 1 (mod.n)$. Together with (2.13) this implies $h_i|d(\gamma_i)\ell_i$. Assertion (2.14) follows.] Now (2.13) and (2.14) yield

$$|\mathbb{Z}_n^* \backslash R_n| \le \prod_{i=1}^{t} \frac{\varphi(q_i)}{2^{\alpha_i} m_i} \le \frac{\varphi(n)}{2^{\alpha} \prod_{i=1}^{t} m_i}.$$

So it suffices to prove that $\beta = 1$. Assume, on the contrary, that $\beta \geq 2$. All e_i with $i \in J$ have the same common value e^*, say. Put $\gamma'_j := f_j + 1$, which also has the same value γ, say, for all $j \in J$. So $h_j/2 | d(\gamma)$, but h_j is not a divisor of $d(\gamma)$. However, due to (2.13), for all $j \in J$ we have

$$x^{d(\gamma)} = 1(mod.q_j) \Longleftrightarrow \varphi(q_j) | \ell_j m_j d(\gamma) \Longleftrightarrow h_j | \ell_j d(\gamma).$$

On the other hand, $\gcd(x^{d(\gamma)} - 1, n) \in \{1, n\}$ since $x \in \mathbb{Z}_n^* \backslash R_n$. So $2 | \ell_j$ is either true for all $j \in J$ or false for all $j \in J$. Now the assertion follows from the fact that $2^{\alpha_i} | \ell_i$ for all $i \in I$. $\square$

The following corollary gives a still more explicit estimate of the primality constant for all relevant cases:

Corollary 2.1. *For all odd composite integers $n \geq 11$, it holds that*

$$\frac{|\mathbb{Z}_n^* \backslash R_n|}{\varphi(n)} \leq \frac{1}{4}.$$

Proof: In case $t \geq 3$, the corollary follows directly from Theorem 2.2. The same is the case if $t = 2$ and either $m_1 = 2$ or $m_2 = 2$ (since $r = 2$ implies $\alpha + \beta - 1 \geq 1$). We consider first the case $t = 2$, $m_1 = m_2 = 1$. So we may write $n = p_1 p_2$, w.l.o.g. with $p_1 < p_2$. But then

$$p_2 - 1 = \varphi(p_2) | n - 1 = p_1(p_2 - 1) + (p_1 - 1),$$

which is not possible. So it remains the case $t = 1$. If we write $n = p^k$ for some $k \geq 2$, we get

$$|\mathbb{Z}_n^* \backslash R_n| \leq |\{x \in \mathbb{Z}_n^* : x^{n-1} = 1(mod.n)\}| \leq \gcd(n - 1, p - 1) = p - 1$$

and thus

$$\frac{|\mathbb{Z}_n^* \backslash R_n|}{\varphi(n)} \leq \frac{p - 1}{p^{k-1}(p - 1)} = \frac{1}{p^{k-1}} \leq \frac{1}{4}$$

since $p \geq 11$. $\square$

For integers n with many different prime factors, we have even a better estimate of the primality constant (see Kranakis (1986), Theorem 2.34):

Corollary 2.2. *For odd integers $n > 2$ whose number of distinct prime factors is t, we have*

$$\frac{|\mathbb{Z}_n^* \backslash R_n|}{\varphi(n)} \leq \frac{1}{2^{t-1}}.$$

For further quantitative results in this context see Darmgård et al. (1993).

2.4 *Bit Security of RSA

Denote by $\mathrm{Lsb}(x)$ the least significant bit of the natural number x (represented in its binary expansion). By a little abuse of notation, we will also write $\mathrm{Lsb}(\overline{x})$ for $\overline{x} := x(mod.n)$ represented as an element of $\{0, 1, \ldots, n-1\}$. As before, let p, q be distinct odd primes, $n := pq$. Assume the RSA exponent e is relatively prime to $\varphi(n)$. The following theorem says that if a polynomial-time algorithm for calculating the least significant bit of the plaintext x exists, then a polynomial-time algorithm for calculating the whole of x also exists. Similar considerations can also be made e.g., for the Rabin system, see Kranakis (1986), 5.7 and also Delfs, Knebl (2002), 7.

Theorem 2.3. *If there exists a polynomial-time algorithm*

$$A_1 = A_1(n, e, y) = \mathrm{Lsb}(x) \quad (x \in \mathbb{Z}_n^*),$$

then there is also a polynomial-time algorithm

$$A_2 = A_2(n, e, y) = x \quad (x \in \mathbb{Z}_n^*).$$

Proof: The method of proof is rational approximation, i.e. to calculate $a \in \mathbb{Z}_n^*$ and $u \in [0, 1[\cap \mathbb{Q}$ such that

$$|\overline{ax} - un| < \frac{1}{2}.$$

The algorithm proceeds recursively. Let $u_0 := 0$ and $a_0 := 1$ be the starting values. Let 2^{-1} denote the inverse of $2(mod.n)$. Define recursively

$$a_t := 2^{-1} a_{t-1}$$

and

$$u_t := \frac{1}{2}(u_{t-1} + \mathrm{Lsb}(\overline{a_{t-1}x})).$$

With this definition, we obtain

$$
\overline{a_t x} = \overline{2^{-1} a_{t-1} x}
= \begin{cases} \frac{1}{2}\overline{a_{t-1}x} & : \quad \overline{a_{t-1}x} = 0(mod.2) \\ \frac{1}{2}(\overline{a_{t-1}x} + n) & : \quad \overline{a_{t-1}x} = 1(mod.2). \end{cases} \tag{2.15}
$$

By the fact that

$$\mathrm{Lsb}(\overline{a_t x}) = A_1(n, e, \overline{a_t^e y})$$

it is possible to decide whether $\overline{a_t x}$ is even from the data available to Eve (i.e., without knowing x explicitly) and thus the recursion step (2.15) can really be done by Eve. So

$$|\overline{a_0 x} - u_0 n| < n,$$

$$|\overline{a_t x} - u_t n| = \frac{1}{2}|\overline{a_{t-1}x} - u_{t-1}n|,$$

and hence after $|n| + 1$ steps (where $|n|$ means the length of the binary expansion of n) she will have

$$|\overline{a_{|n|+1}x} - u_{|n|+1}n| < \frac{n}{2^{|n|+1}} < \frac{1}{2}. \qquad (2.16)$$

But from (2.16) it follows that

$$\overline{a_{|n|+1}x} = \lfloor u_{|n|+1}n + \frac{1}{2} \rfloor$$

and hence

$$x = a_{|n|+1}^{-1} \lfloor u_{|n|+1}n + \frac{1}{2} \rfloor (mod.n). \square$$

An analogue of Theorem 2.3 also exists for probabilistic algorithms.

Definition 2.1. *A probabilitstic algorithm is an algorithm A that, during the computation of the output y from the input x, is allowed to generate a finite number of independent unbiased random bits, and the next step may depend on the results of the preceding random bits. The number of random bits may depend on the outcome of the previous ones, but is bounded by some constant t_x for a given input x.*
A probabilistic algorithm is called polynomial-time (or polynomial) if the running time of $A(x)$ is bounded by some polynomial $\xi(|z|)$ that is independent of z. (Generating a random bit counts as one step in the complexity of the algorithm.)

A polynomial $\xi(z)$ is called positive, if $\xi(z) > 0$ for all $z > 0$. The following theorem is the probabilistic analogue of Theorem 2.3:

Theorem 2.4. *Let p, q be distinct odd primes and write $n := pq$ for their product. Assume e is relatively prime to $\varphi(n)$ and denote $y := x^e (mod.n)$. Let ξ and η be positive polynomials with integer coeffficients. Suppose there exists a probabilistic polynomial time algorithm A_1 such that, for uniformly distributed x on $\mathbb{Z}_n^*$, it holds that*

$$P(A_1(n, e, y) = \mathrm{Lsb}(x)) \geq \frac{1}{2} + \frac{1}{\xi(|n|)}.$$

Then there exists a polynomial-time algorithm A_2 such that

$$P(A_2(n, e, y) = x) \geq 1 - 2^{-\eta(|n|)}.$$

The proof of Theorem 2.4 rests on the following lemmas. The first one is just a consequence of a quantitative version of the Weak Law of Large Numbers:

Lemma 2.5. *Assume $S_1, S_2, \ldots, S_n$ are pairwise independent binary random variables with common expectations $E(S_i) =: \alpha = \frac{1}{2} + \varepsilon$ ($\varepsilon > 0$). Then*

$$P(\sum_{i=1}^{t} S_i > \frac{t}{2}) \geq 1 - \frac{1}{t\varepsilon^2}.$$

Proof: Observe that

$$|\frac{1}{t}\sum_{i=1}^{t}S_i - \alpha| < \varepsilon$$

implies

$$\frac{1}{t}\sum_{i=1}^{t}S_i > \frac{1}{2},$$

and therefore (with the aid of Čebyšev's inequality and some straightforward calculations) we obtain

$$P(\sum_{i=1}^{t}S_i > \frac{t}{2}) \geq P(|\frac{1}{t}\sum_{i=1}^{t}S_i - \alpha| < \varepsilon)$$

$$\geq 1 - \frac{1}{t\varepsilon^2}.\square$$

Lemma 2.6. *Under the hypotheses of Theorem 2.4, there exists a probabilistic polynomial-time algorithm L with the following properties: If a, b are independent randomly chosen elements of $\mathbb{Z}_n^*$ (according to the uniform distribution on this set), if we take $u, v \in \mathbb{Q}$ such that*

$$|\overline{ax} - un| \leq \frac{\varepsilon^3 n}{8}$$

and

$$|\overline{bx} - vn| \leq \frac{\varepsilon n}{8}$$

(for some $\varepsilon > 0$ small enough), and if we put (recursively) $a_0 := a$ and $a_t := 2^{-1}a_{t-1}$, then L successively computes values ℓ_t (for $t = 0, 1, \ldots, |n|$) such that

$$P(\ell_t = \mathrm{Lsb}(\overline{a_t x}) \mid \ell_j = \mathrm{Lsb}(\overline{a_j x})(0 \leq j \leq t-1)) \geq 1 - \frac{1}{2|n|}. \quad (2.17)$$

(In fact, we choose $a, b \in \mathbb{Z}_n$. But otherwise, then we may factor n just by the Euclidean algorithm.)

Proof of Lemma 2.6: Put $m := \min\{2^t/\varepsilon^2, 2|n|/\varepsilon^2\}$. Then w.l.o.g. we may assume that $p, q > m$ because otherwise, we can factorize n in polynomial time just by exhaustive search.

Put first

$$\alpha := \mathrm{Lsb}(\overline{ax}),$$

$$\beta := \mathrm{Lsb}(\overline{bx}).$$

We now show first how to calculate $\ell_t = \mathrm{Lsb}(\overline{(a_t + ia_{t-1} + b)x})$ (w.l.o.g. we may assume that $a_t + ia_{t-1} + b$ is really invertible mod.n, for otherwise we can factor n just with the Euclidean algorithm). The following subroutine (which calculates ℓ_t, a_t, and u_t recursively (the resulting algorithm will be called L')) is run: The initial value is $\ell_0 := \alpha$, $a_{t-1} := a_0 := a$, and $u_{t-1} := u$:

- $C_0 \leftarrow 0; C_1 \leftarrow 0;$
- $a_t \leftarrow 2^{-1} a_{t-1}; u_t \leftarrow \frac{1}{2}(u_{t-1} + \alpha);$
- FOR $i = -m/2$ to $m/2 - 1$ DO
  - $A \leftarrow a_t + i a_{t-1} + b;$
  - $W \leftarrow \lfloor u_t + i u_{t-1} + v \rfloor;$
  - $B \leftarrow (i\alpha + \beta + W)(mod.2);$
  - IF $A_1(n, e, A^e y (mod.n)) + B = 0$
    - THEN $C_0 \leftarrow C_0 + 1;$
    - ELSE $C_1 \leftarrow C_1 + 1,$
- $u_{t-1} \leftarrow u_t; a_{t-1} \leftarrow a_t;$
- IF $C_0 > C_1$
  - $\alpha \leftarrow 0;$
  - $\alpha \leftarrow 1;$
- RETURN $\alpha;$

So we have got the "modified value" $\alpha := \ell_t = \mathrm{Lsb}(\overline{(a_t + i a_{t-1} + b)x})$.
Now we will calculate what we really want, namely $\mathrm{Lsb}(\overline{a_t x})$. We will see that the hypotheses of Lemma 2.6 guarantee that we can indeed infer $\mathrm{Lsb}(\overline{a_t x})$ with high probability. For $i = -m/2, -m/2+1, \ldots, m/2 - 1$ define

$$A_{t,i} := a_t + i a_{t-1} + b,$$

$$W'_{t,i} := u_t + i u_{t-1} + v,$$

$$W_{t,i} := \lfloor W'_{t,i} \rfloor,$$

$$B_{t,i} := (i \cdot \mathrm{Lsb}(\overline{a_{t-1}x}) + \mathrm{Lsb}(\overline{bx}) + \mathrm{Lsb}(W_{t,i}))(mod.2).$$

We want to compute $\mathrm{Lsb}(\overline{a_t x})$ (recursively) from the data

$$\mathrm{Lsb}(\overline{A_{t,i}x}), \quad \mathrm{Lsb}(\overline{a_{t-1}x}), \quad \mathrm{Lsb}(\overline{bx}).$$

Put

$$\lambda_{t,i} := \overline{a_t x} + i \cdot \overline{a_{t-1}x} + \overline{bx}$$
$$= wn + \overline{A_{t-i}x}$$

where

$$w := \lfloor \lambda_{t,i}/n \rfloor.$$

Then

$$\mathrm{Lsb}(\lambda_{t,i}) = (\mathrm{Lsb}(\overline{a_t x}) + i \cdot \mathrm{Lsb}(\overline{a_{t-1}x}) + \mathrm{Lsb}(\overline{bx}))(mod.2)$$

and

$$\mathrm{Lsb}(\overline{A_{t,i}x}) = (\mathrm{Lsb}(\lambda_{t,i}) + \mathrm{Lsb}(w))(mod.2)$$
$$= (\mathrm{Lsb}(\overline{a_t x}) + i \cdot \mathrm{Lsb}(\overline{a_{t-1}x}) + \mathrm{Lsb}(\overline{bx}) + \mathrm{Lsb}(w))(mod.2),$$

and we obtain

$$\mathrm{Lsb}(\overline{a_t x}) = (\mathrm{Lsb}(\overline{A_{t,i} x}) + i \cdot \mathrm{Lsb}(\overline{a_{t-1} x}) + \mathrm{Lsb}(\overline{bx}) + \mathrm{Lsb}(w))(mod.2).$$

Now let us determine w and its least significant bit $\mathrm{Lsb}(w)$. The method will be to show that w equals $W_{t,i}$ with high probability and that, on the other hand, it is really possible to compute $W_{t,i}$ in polynomial time from the available data u_t (the rational approximation of $\overline{a_t x}$), u_{t-1} (the rational approximation of $\overline{a_{t-1} x}$), and v (the rational approximation of $\overline{bx}$). If indeed $W_{t,i} = w$, we have

$$\mathrm{Lsb}(\overline{a_t x}) = (\mathrm{Lsb}(\overline{A_{t,i} x}) + B_{t,i})(mod.2).$$

Now assume that the algorithm L' has computed the least significant bit correctly in all preceeding steps, i.e.,

$$\mathrm{Lsb}(\overline{a_j x}) = \ell_j \quad (0 \le j \le t-1).$$

We intend to give a lower bound for the probability that $W_{t,i} = w$. Denote the random variable

$$Z := |\lambda_{t,i} - W'_{t,i} n|.$$

We may estimate

$$
\begin{aligned}
Z &= |\overline{a_t x} - u_t n + i(\overline{a_{t-1} x} - u_{t-1} n) + \overline{bx} - vn| \\
&\le |\frac{1}{2}(\overline{a_{t-1} x} + u_{t-1} n)(1 + 2i)| + |\overline{bx} - vn| \\
&\le \frac{n}{2^t} \frac{\varepsilon^3}{8} |1 + 2i| + \frac{\varepsilon}{8} n \\
&\le \frac{\varepsilon}{8} n (\frac{\varepsilon^2 m}{2^t} + 1) \\
&\le \frac{\varepsilon}{4} n. \quad\quad\quad (2.18)
\end{aligned}
$$

Under our assumption that $\ell_j = \mathrm{Lsb}(\overline{a_j x})$ $(j = 0, 1, \ldots t-1)$ if follows (as in the proof of Theorem 2.3) that

$$|\overline{a_j x} - u_j n| = \frac{1}{2} |(\overline{a_{j-1} x} - u_{j-1} n)| \quad (1 \le j \le t).$$

Furthermore $|1 + 2i| \le m$ (since $-m/2 \le i \le m/2 - 1$). Now we observe that $W_{t,i} \ne w$ iff there is a multiple of n between $\lambda_{t,i}$ and $W'_{t,i} n$. The latter is not the case, if the following holds:

$$\frac{\varepsilon}{4} n < \overline{\lambda_{t,i}} = \overline{A_{t,i} x} < n - \frac{\varepsilon}{4} n.$$

Hence by the uniform distribution of a and b on $\mathbb{Z}_n^*$ it follows that the

$$\overline{\lambda_{t,i}} = \overline{(2^{-1}+i)a_{t-1} + bx}$$

are also uniformly distributed and thus

$$P(W_{i,t} = w) \geq P(\frac{\varepsilon}{4}n < \overline{A_{t,i}x} < n - \frac{\varepsilon}{4}n)$$

$$\geq 1 - \frac{\varepsilon}{2}.$$

Now we want to show

$$P(\ell_t = \mathrm{Lsb}(\overline{a_t x}) \mid \ell_j = \mathrm{Lsb}(\overline{a_j x})(0 \leq j \leq t-1)) \geq 1 - \frac{1}{2|n|}. \qquad (2.19)$$

Consider the events

$$E_{1,i} := \{A_1(n, e, \overline{A_{t,i}^e y}) = \mathrm{Lsb}(\overline{A_{t,i}x})\}$$

and

$$E_{2,i} := \{\frac{\varepsilon}{4}n < \overline{A_{t,i}x} < n - \frac{\varepsilon}{4}n\}.$$

It follows that $P(E_{1,i}) \geq \frac{1}{2} + \varepsilon$ and $P(E_{2,i}) = 1 - \varepsilon/2$. Consider the indicator random variables $I_i := 1_{(}E_{1,i} \cap E_{2,i})$. The algorithm L computes $\mathrm{Lsb}(a_t x)$ correctly in the i-th step if both events $E_{1,i}$ and $E_{2,i}$ occur. So it follows that

$$P(I_i = 1) \geq P(E_{2,i}) - (1 - P(E_{1,i}))$$

$$> (1 - \frac{\varepsilon}{2}) - (\frac{1}{2} - \varepsilon)$$

$$= \frac{1}{2} + \frac{\varepsilon}{2}.$$

Now assume $i \neq j$. Take the probabilities $P(I_i = d)$ and $P(I_j = d)$ ($d \in \{0,1\}$) over all random choices of $a, b \in \mathbb{Z}_n^*$ and the random bit generations produced by the algorithms $A_1(n, e, \overline{A_{t,i}^e y})$ and $A_1(n, e, \overline{A_{t,j}^e y})$. If we define the 2×2-matrix

$$\Delta := \begin{pmatrix} 2^{-1}+i & 2^{-1}+j \\ 1 & 1 \end{pmatrix}$$

which is invertible over $\mathbb{Z}_n^*$ and has determinant $i - j \in \mathbb{Z}_n^*$ (since $|i - j| < m < \min\{p, q\}$), then we have

$$(\overline{A_{t,i}}, \overline{A_{t,j}}) = (a_{t-1}, b)\Delta = (2^{-t+1}a, b)\Delta.$$

This implies that for $i \neq j$, the random vectors $\overline{A_{t,i}}$ and $\overline{A_{t,j}}$ are independent. So the events $E_{2,i}$ and $E_{2,j}$ and the random variables $\overline{A_{t,i}^e y}$ and $\overline{A_{t,j}^e y}$ ($i \neq j$) are independent. Hence (for $i \neq j$) the events $E_{1,i}$ and $E_{1,j}$ and thus the indicator variables I_i and I_j are independent. By Lemma 2.5, it follows that

$$P\left(\sum_{i=-m/2}^{m/2-1} I_i > \frac{m}{2}\right) \geq 1 - \frac{1}{n\varepsilon^2} \geq 1 - \frac{1}{2|n|}.$$

If $\mathrm{Lsb}(\overline{a_t x}) = 0$, then we have $C_0 \geq \sum_i S_i$ and thus

$$P(C_0 > C_1) \geq 1 - \frac{1}{2|n|}.$$

On the other hand, if $\mathrm{Lsb}(\overline{a_t x}) = 1$, then by analogy

$$P(C_1 > C_0) \geq 1 - \frac{1}{2|n|}.$$

The assertion of the lemma follows.$\square$
Now we are ready to prove Theorem 2.4:
Proof of Theorem 2.4: We run the following algorithm:

- Choose $a, b \in \mathbb{Z}_n^*$ at random.
- Guess $u, v \in [0, 1] \cap \mathbb{Q}$ such that

$$|\overline{ax} - un| \leq \frac{\varepsilon^3}{8} n$$

and

$$|\overline{bx} - vn| \leq \frac{\varepsilon}{8} n.$$

- Guess $\alpha := \mathrm{Lsb}(\overline{ax}), \beta := \mathrm{Lsb}(\overline{bx})$.
- Compute $\ell_0, \ell_1, \ldots, \ell_{|n|}$ by the algorithm L from Lemma 2.6.
- FOR $t = 0, 1, \ldots, |n|$ DO $u \leftarrow \frac{1}{2}(u + \ell_t), a \leftarrow \overline{2^{-1}a}$.
  RETURN $\overline{a^{-1}\lfloor un + \frac{1}{2}\rfloor}$.

It is easy to see that this algorithm is indeed polynomial, since there are only polynomially many alternatives for all guesses, and both the calculation of each alternative and checking the result can be done in polynomial time. For guessing u, v, one has to consider polynomially many (namely $8/\varepsilon^3$ and $8/\varepsilon$) intervals and there are only 2 possible values for α and β. This algorithm (called A) has success probabilty

$$P(A(n, e, y) = x) \geq \left(1 - \frac{1}{2|n|}\right)^n.$$

If we repeat A sufficiently many times (with independent inputs and every time using the trivial deterministic algorithm for testing the result), the assertion of Theorem 2.4 follows. [The probabilty of a wrong answer in t repetitions is bounded by

$$(\frac{1}{2} - \frac{1}{\xi(|n|)})^t < (1 - \frac{2}{\xi(|n|)})^t$$

$$= (1 - (\frac{2}{\xi(|n|)})^{\xi(|n|)/2})^{2t/\xi(|n|)}$$

$$< e^{-2t/\xi(|n|)}$$

$$\le e^{-\log 2 \cdot \eta(|n|)}$$

$$= 2^{-\eta(|n|)}$$

for large enough t .]$\square$

2.5 The Timing Attack on RSA

The fact that factoring is probably computationally difficult should not lead us to believe that there are no attacks possible on RSA. See, e.g., Boneh (1999). Here, we will present a type of attack based rather on the implementation of RSA than on the algorithm itself. It may be possible for Eve to measure the time a smartcard uses for performing RSA-operations. With this, she may be able to recover the private key d_A of Alice. We first show the repeated squaring algorithm for computing $y = x^{d_A} (mod.n_A)$, which runs in time linear in $\log d_A$. Let

$$d_A = d_m d_{m-1} \ldots d_0$$

be the binary expansion of d_A. Observe that

$$y = \prod_{i=0}^{m} x^{2^i d_i} (mod.n_A).$$

The repeated squaring algorithm works as follows:

- Put the initial values $X \leftarrow x$ and $Y \leftarrow 1$.
- For $i = 0, 1, \ldots, n$ put $Y \leftarrow Y \cdot X(mod.n_A)$ (if $d_i = 1$) and $X \leftarrow X^2(mod.n_A)$.

Then at the end we have $Y = y$.

The timing attack can now be mounted by Eve as follows: She takes a large number k of random plaintexts $x_1, x_2, \ldots, x_k \in \mathbb{Z}_{n_A}^*$ and measures the time T_i the smartcard uses for encrypting x_i. Now she may recover the bits d_i of d_A in the following way: Of course d_A is odd, so $d_0 = 1$. In the second iteration, the smartcard computes $Y \cdot X = X \cdot X^2(mod.n_A)$ iff $d_1 = 1$. Let t_i denote the time the smartcard uses for computing $x_i \cdot x_i^2(mod.n_A)$. Eve can have measured these t_i offline before mounting the attack and compares them now with the T_i. Namely, it turns out (Kocher) that if $d_1 = 1$, the sequences $\{T_i\}_{1 \le i \le k}$ and $\{t_i\}_{1 \le i \le k}$ are (positively) correlated, whereas in the other case they behave as independent random variables. So by measuring the correlation of $\{T_i\}_{1 \le i \le k}$ and $\{t_i\}_{1 \le i \le k}$, Eve can guess d_1, etc.

Of course, this attack can be avoided by adding artificial delays on the smart-card so that all modular exponentiations take the same time and can there-fore not be distinguished by measuring time. Another possibility is so-called blinding, suggested by Rivest: Before encrypting the plaintext x, one picks a random $r \in \mathbb{Z}_{n_A}^*$ and replaces x by

$$x' := xr^{e_A}(mod.n_A).$$

Now the RSA-encryption of x' yields

$$y' = (x')^{d_A}(mod.n_A)$$

and the output of the smartcard is

$$y = y'/r(mod.n_A).$$

Here, the RSA-exponentiation with d_A has been applied to x' that behaves randomly, so the timing attack before is not possible.

If the exponentiation with the secret exponent is done by Montgomery's multiplication algorithm for the prime factors of the secret exponent and the Chinese remainder theorem to obtain the final result, then one can not use the above-mentioned attack, but rather the one described in Schindler (2000). Other related attacks are the measuring of power consumption of the smart-card. As a consequence one sees that despite the theoretical strength of the RSA system, its implementation in hardware must be done with much care. (This is, of course, also true for other cryptosystems.) For more information on timing attacks, see e.g. Schindler (2002a). Here, interesting methods from statistical desicion theory (which are beyond the scope of this text) come into play. A combined timing and power attack on RSA was presented in Schindler (2002b) and Schindler, Walter (2003). An interesting observation in this direction for elliptic-curve cryptosystems was made by Okeya, Sakurai (2000).

2.6 *Zero-Knowledge Proof for the RSA Secret Key

Up to now, we have always assumed that the parties Alice and Bob trust each other and that they only want to prevent Eve from eavesdropping. In this short section, we take another view: Alice wants to convince Bob that she knows some secret but she does not want to give Bob any information about the secret itself. For example, she wants to tell Bob that she knows, say, Carol's private RSA key, but she does not want to give him any hint as to the key itself or even decrypt one of Carol's messages to Bob. Let $e = e_C$ resp. $d = d_C$ be Carol's private resp. public key. The zero-knowledge protocol will involve an interactive fair so-called coin-flipping subprotocol. There are different ways to do this. We present the method with the square roots. Other methods use exponention modulo a prime or Blum integers (see Schneier (1996), pp. 542f.)

1. Alice chooses two large primes p, q and sends their product $n := pq$ to Bob.
2. Bob chooses a random natural number $r < n/2$ and sends

$$z := r^2 (mod.n)$$

   to Alice.
3. Since she knows p and q, Alice can compute the 4 square roots $x, -x, y, -y$ (say) of $z(mod.n)$. Denote (with slight abuse of notation)

$$x' := \min\{x(mod.n), -x(mod.n)\}$$

   and

$$y' := \min\{y(mod.n), -y(mod.n)\}.$$

   Then we have $r \in \{x', y'\}$.
4. Alice guesses if $r = x'$ or $r = y'$ and transmits her guess to Bob.
5. If Alice's guess was correct, then the coin-flipping subalgorithm outputs 1, otherwise 0.
6. Verification subsubprotocol: Alice sends p and q to Bob, Bob computes x' and y' and sends them to Alice, then Alice calculates r.

Since Alice can not know r, her guess is really random. In step 4, she tells Bob only one bit of her guess in order to prevent him from obtaining both x' and y'. If Bob has both of these two numbers, he can change r after step 4. Now the zero-knowledge protocol proceeds as follows:

– Alice and Bob agree on random k, m such that

$$km = e(mod.n)$$

   (with $k, m \geq 3$, otherwise they restart the algorithm) using a coin-flipping protocol.
– Again by a coin-flipping protocol, Alice and Bob generate a random ciphertext y.
– Alice uses Carol's private key to compute

$$x = y^d (mod.n)$$

   and

$$t := x^k (mod.n)$$

   and sends t to Bob.
– Bob checks if $t^m = y(mod.n)$. If yes, he believes Alice.

This protocol can be rerun several times. Then the probability that Alice bluffs decreases exponentially with the number of times the algorithm is executed.

3 Factorization with Quantum Computers: Shor's Algorithm

3.1 Classical Factorization Algorithms

The most famous classical factorization algorithms are the Quadratic Sieve (QS) and the Number Field Sieve (NFS). Though being subexponential, they are not polynomial. The QS is the fastest general-purpose factorization algorithm for numbers with less than 110 digits, whereas the NFS has the same property for numbers with more than 110 digits (see Schneier (1996), p.256). Recently, RSA-576 with a number of 174 decimal digits was factorized by Franke from the Universoty of Bonn with the aid of the NFS. The NFS was also used to factorize the Mersenne number $2^{757} - 1$ (with 288 decimal digits) by the Internet project NESNET (about 5 months of computing time on up to 120 machines was necessary). Further limits on the factorization of large numbers can be found on the Internet site CiteSeer. For particular types of numbers to be factorized, many specially designed algorithms have been developed, which in these cases are faster than the above-mentioned ones. A new direction of cryptanalysis would be the possible use of quantum computers istead of classical Turing machines. Up to now, quantum computing has been more or less only a theoretical concept based on the superposition principle of quantum mechanics. Beyond some basic experiments, nobody has really an idea how to realize physical quantum computers working efficiently. However, if one day quantum computers could be built, this would have dramatic consequences for cryptology. Namely, in the second half of the 1990s, Peter Shor showed that on a quantum computer, large numbers can be factorized in linear (with respect to the length of the binary expansion of the number) time! So in this case, the RSA and all related systems would be worthless against an adversary who has a quantum computer at his disposal. Shor's method is in fact a hybrid algorithm in the sense that it consists of four components, one being done by quantum computing and three others (a little trick based on the Euclidean algorithm from elementary number theory, Fourier transform and continued-fraction approximation) that can be done on a classical computer. (Note that the Fourier transform component can, but need not be done on a quantum computer.) This algorithm will be explained in Section 3.4. We note that Shor has developed another algorithm for solving the discrete logarithm problem on quantum computers. Here, we will not discuss that, but the principles are similar.

D. Neuenschwander: Prob. and Stat. Methods in Cryptology, LNCS 3028, pp. 37-45, 2004.
© Springer-Verlag Berlin Heidelberg 2004

Note that in contrast to Chapter 13, where we will present the ideas of quantum cryptography, here we use quantum computers to cryptanalyze classical cryptosystems.

We remark that there is a new approach due to Hungerbühler, Struwe (2003), who suggest the heat flow as a cryptographic system that resists also attacks by quantum computers. It is based on the second principle of thermodynamics (increase of entropy). The evolution problem for the heat equation is a well-posed initial-value problem, which can be solved very precisely by numerical methods, whereas the evolution problem in backward time is ill-posed and numerical methods for solving the heat equation for negative time are inherently unstable.

3.2 Quantum Computing

Let us now give a short introduction to quantum computing, which rests on a non-Kolmogorovian type of probability, namely quantum stochastics. In the following, we will present some basic facts on quantum mechanics and quantum computing. In quantum physics, the state of a quantum system is described by a vector in a (complex) Hilbert space H. It is customary to write such a state vector as a column vector, or - in the jargon of quantum physics - as "*ket* vector" $|\psi\rangle$. The corresponding line vector is written as $\langle\psi|$ and called "*bra* vector". The squared norm of the vector, or - in other words - the scalar product of the vector with itself is then written as $\langle\psi|\psi\rangle$, which becomes a *bracket*. We now come to the process of measurement in quantum mechanics. As a principle, in quantum mechanics measurements of observables are described by Hermitian operators A acting on the underlying Hilbert space H. If the system is in an eigenstate of A, then the measurement with the operator A just reproduces the state, multiplied with a real number (since A is Hermitean). If the system is not in an eigenstate, then the outcome of the measurement will collapse to one of the observables (corresponding to eigenstates (eigenvalues)) of A, but what is important is that it cannot be predicted in advance to which one. Only probabilites can be indicated, which correspond to the principle of superposition. The result of any measurement of a quantum system described by the state vector $|\psi\rangle$ is always one of the eigenvalues of the operator A, corresponding to the observable being measured. If the system is in an eigenstate of A, then the outcome of the measurement is just the corresponding eigenvalue of this eigenstate. In general, the system will be in some general state ϕ. Then we may represent ϕ as a complex linear combination with respect to a basis $\{\psi_i\}_i$ of eigenstates of A:

$$|\phi\rangle = \sum_i \omega_i |\psi_i\rangle, \tag{3.1}$$

where the ω_i are called the probability amplitudes. If w.l.o.g. we assume that $\sum_i |\omega_i|^2 = 1$, then $|\omega_i|^2$ is interpreted as the probability that the system is in eigenstate i with respect to A. So a quantum system can exist in a blend of all its eigenstates with respect to a certain Hermitian measurement operator simultaneously. This is called the principle of superposition and is the big difference between quantum and classical mechanics, where absolutely no such analogue exists. If the system is in a superposition of states as in (3.1), then the probability of each possible outcome of the measurement A (i.e. of each possible eigenvalue) is given by $|\omega_i|^2$. An unobserved quantum system is governed by Schrödinger's equation

$$ih|\dot{\psi}(t)\rangle = \hat{H}(t)|\psi(t)\rangle,$$

where $h = 1.0545 \cdot 10^{-34} Js$ is Planck's constant and $\hat{H}(t)$ is the Hamiltonian (unitary operator) related to the total energy of the system; so the system behaves smoothly until it is measured.

Definition 3.1. *A qubit (quantum bit) is a quantum 2-state system*

$$|\psi\rangle = \alpha|0\rangle + \beta|1\rangle, \tag{3.2}$$

where $\alpha, \beta \in \mathbb{C}$ such that $|\alpha|^2 + |\beta|^2 = 1$. (Note that $|0\rangle$ and $|1\rangle$ are just names for the eigenvectors representing a classical bit and have nothing to do with the zero vector in the Hilbert space $H = \mathbb{C}^2$.)

It is an easy exercise to show that for $|\psi\rangle$ as in (3.2) there exist angles θ, ϕ such that

$$|\psi\rangle = \cos\theta|0\rangle + e^{i\phi}\sin\theta|1\rangle. \tag{3.3}$$

So a (single) qubit can, geometrically, be interpreted as a point on the two-dimensional unit sphere, the north pole (e.g.) representing the eigenstate $|1\rangle$ and the south pole the eigenstate $|0\rangle$. It turns out that information that in classical computers use much memory can be stored with much fewer qubits in quantum computers. Let us begin with a simple example: Assume we have two classical complementary bitstrings of length 7, e.g., $|0110101\rangle$ and $|1001010\rangle$. In order to store them in a classical computer, we need two registers each of length 7 bit. However, on a quantum computer, only one register of 7 qubits suffices, since here we can just store the superposition $\frac{1}{\sqrt{2}}(|0110101\rangle + |1001010\rangle)$. More generally, if we have an exponential number of bits to store, by using the superposition principle, a polynomial number of qubits suffices. (Of course, these superpositions can be very complicated in general.) An n-qubit memory register is realized by the n-fold tensor product of the 1-qubit register. However, quantum evolvement of such a register may lead to states that are defined as a whole, but do not arise from individual qubits, i.e., the individual qubits are not defined as such. Such states are called entangled (Schrödinger used the german word "verschränkt" for it). A very important fact is that measurements of subsets of qubits in an n-qubit register project out the state of the whole register into a subset of eigenstates consistent with the answers (eigenvalues) obtained from the measurement.

This has to do with quantum teleportation and the Einstein-Podolsky-Rosen experiment. Another aspect is quantum parallelism. Quantum evolution is performed by unitary operators (on "single processors"), which operate at the same time on all possible states. These phenomena will be crucial in Shor's algorithm.

3.3 Continued Fractions

A regular continued fraction is an expression of the type

$$a_0 + \cfrac{1}{a_1 + \cfrac{1}{a_2 + \cfrac{1}{a_3 + \cfrac{1}{a_4 \dots}}}} \qquad (3.4)$$

with finitely or infinitely many members $a_k \in I\!N$ ($k \in I\!N$), $a_0 \in Z\!\!\!Z$. For graphical simplicity, it is also customary to write (3.4) as

$$[a_0, a_1, a_2, a_3, a_4, \dots]. \qquad (3.5)$$

For an infinite regular continued fraction (3.5), finite expressions

$$[a_0, a_1, \dots, a_n]$$

are called "approximating fractions". Finite regular continued fractions represent rational numbers (moreover, the a_k are uniquely determined if we suppose that the last one of them is ≥ 2), whereas for irrational numbers the following Theorem 3.1 on "continued fraction approximation" is valid. Often we will use the notation

$$\xi = [a_0, a_1, \dots]$$

for both finite and infinte continued fractions (i.e. for both ξ rational or irrational). When ξ is rational, then the above will mean that finite continued fraction whose last denominatior is > 1.

Theorem 3.1. *Every infinite regular continued fraction $[a_0, a_1, \dots]$ converges to an irrational number ξ_0. Conversely, every irrational number ξ_0 is the limit of a unique regular continued fraction, which is necessarily infinite:* $\xi_0 = [a_0, a_1, \dots]$.

Proof: 1. It is easily verified that one may write

$$[a_0, a_1, \dots, a_n] = \frac{A_n}{B_n}$$

if one defines the A_n and B_n by the linear depth 2-recursion $A_{-1} := 1, B_{-1} := 0, A_0 := a_0, B_0 := 1,$

$$A_n := a_n A_{n-1} + A_{n-2},$$
$$B_n := a_n B_{n-1} + B_{n-2} \quad (n \geq 1). \tag{3.6}$$

It follows from (3.6) that $\lim_{n \to \infty} A_n = \lim_{n \to \infty} B_n = \infty$. Consider the system of equations

$$\xi_n = a_n + \frac{1}{\xi_{n+1}} \quad (n \in \mathbb{N}_0). \tag{3.7}$$

System (3.7) can also be written as

$$\xi_0 = [a_0, a_1, \ldots, a_{n-1}, \xi_n] \quad (n \in \mathbb{N}), \tag{3.8}$$

which is equivalent to

$$\cdot \quad \xi_0 = \frac{A_{n-1}\xi_n + A_{n-2}}{B_{n-1}\xi_n + B_{n-2}}. \tag{3.9}$$

Hence we get

$$\xi_0 - \frac{A_{n-1}}{B_{n-1}} = \frac{A_{n-2}B_{n-1} - A_{n-1}B_{n-2}}{B_{n-1}(B_{n-1}\xi_n + B_{n-2})} = \frac{(-1)^{n-1}}{B_{n-1}(B_{n-1}\xi_n + B_{n-2})} \to 0$$
$$(n \to \infty), \tag{3.10}$$

i.e. indeed

$$\xi_0 = \lim_{n \to \infty} \frac{A_n}{B_n} = [a_0, a_1, \ldots]. \tag{3.11}$$

2. Now we show that every infinite regular continued fraction converges to some limit ξ_0, i.e., that

$$\xi_0 := \lim_{n \to \infty} \frac{A_n}{B_n} \tag{3.12}$$

exists. For the difference between two consecutive approximating fractions, after some calculations, one obtains the estimate

$$\left| \frac{A_{n+m-1}}{B_{n+m-1}} - \frac{A_{m-1}}{B_{m-1}} \right| < \frac{1}{B_{m-2}B_{m-1}}, \tag{3.13}$$

which shows that the sequence $\{\frac{A_n}{B_n}\}_{n \geq 1}$ is indeed a Cauchy sequence, i.e. convergent to some real number ξ_0.

3. It remains to prove that the ξ_0 as it was just defined in 2. is irrational. Put

$$\xi_n := [a_n, a_{n+1}, \ldots] \quad (n \in \mathbb{N}_0). \tag{3.14}$$

From (3.9), this yields (similarly as for (3.10))

$$B_{n-1}\xi_0 - A_{n-1} = \frac{B_{n-1}A_{n-2} - B_{n-2}A_{n-1}}{B_{n-1}\xi_n + B_{n-2}} = \frac{(-1)^{n-1}}{B_{n-1}\xi_n + B_{n-2}} \to 0$$
$$(n \to \infty). \tag{3.15}$$

So, as $n \to \infty$, the expression $B_{n-1}\xi_0 - A_{n-1}$ tends to zero without really becoming zero, which is only possible if ξ_0 is irrational (for, otherwise, if $\xi_0 = p/q$ ($p, q \in \mathbb{Z}$), the term $|B_{n-1}\xi_0 - A_{n-1}| = |B_{n-1}p - A_{n-1}q|/|q|$ cannot fall below $1/|q|$ without becoming zero).

4. Eventually, we show that the approximation of an irrational number by an infinite regular continued fraction is unique. Of course we have $\xi_n \geq a_n$ for all n. Since ξ_0 and thus all ξ_n are irrational, equality is not possible, so we have $\xi_n > a_n$ for all $n \in \mathbb{N}_0$. Now if

$$\xi_0 = [b_0, b_1, \ldots]$$

and if we put

$$\eta_n := [b_n, b_{n+1}, \ldots],$$

then we also have

$$\eta_{n-1} = [b_{n-1}, b_{n-2}, \ldots]$$

and thus

$$\eta_{n-1} = [b_{n-1}, \eta_n] = b_{n-1} + \frac{1}{\eta_n}. \tag{3.16}$$

Since $\eta_n > b_n \geq 1$, it follows that b_{n-1} is the largest integer contained in η_{n-1}, so in particular, b_0 is the largest integer contained in η_0, and thus $b_0 = a_0$. But in this case, equation (3.16) yields η_1, and by the same reasoning as before, it follows that $b_1 = a_1$, etc. $\square$

The following proposition (whose proof is just a short verification) will be of importance in the proof of Theorem 3.2:

Proposition 3.1. *If*

$$\xi_0 = [a_0, a_1, \ldots, a_{n-1}, \xi_n]$$

and

$$\xi_n = [a_n, a_{n+1}, \ldots],$$

then it follows that

$$\xi_0 = [a_0, a_1, \ldots, a_n, a_{n+1}, \ldots].$$

Now we are ready to state the result that will be used to develop Shor's algorithm:

Theorem 3.2. *If $c, d \in \mathbb{Z}$ obey the inequality*

$$\left|\xi_0 - \frac{c}{d}\right| < \frac{1}{2d^2}, \tag{3.17}$$

then $\frac{c}{d}$ is an approximating fraction of ξ_0.

Proof: Put

$$\xi_0 - \frac{c}{d} = \frac{\theta\delta}{d^2}, \tag{3.18}$$

where $\delta = \pm 1$ and then $0 < \theta < \frac{1}{2}$. Furthermore, we write $\frac{c}{d}$ as a finite continued fraction

$$\frac{c}{d} =: [a_0, a_1, \ldots, a_{n-1}], \tag{3.19}$$

where we choose n even or odd such that $(-1)^{n-1} = \delta$. If we define A_k and B_k $(0 \le k \le n-1)$ as in the proof of Theorem 3.1 (so in particular $A_{n-1} = c$ and $B_{n-1} = d$) and ω by the equation

$$\xi_0 =: \frac{A_{n-1}\omega + A_{n-2}}{B_{n-1}\omega + B_{n-2}}, \tag{3.20}$$

then this is equivalent to

$$\frac{\delta\theta}{B_{n-1}^2} = \xi_0 - \frac{A_{n-1}}{B_{n-1}} = \frac{A_{n-2}B_{n-1} - A_{n-1}B_{n-2}}{B_{n-1}(B_{n-1}\omega + B_{n-2})} = \frac{(-1)^{n-1}}{B_{n-1}(B_{n-1}\omega + B_{n-2})} \tag{3.21}$$

or - in other words -

$$\omega = \frac{B_{n-1} - \theta B_{n-2}}{\theta B_{n-1}} > 0. \tag{3.22}$$

Equation (3.20) may be rewritten as

$$\xi_0 = [a_0, a_1, \ldots, a_{n-1}, \omega]. \tag{3.23}$$

Since $\theta < \frac{1}{2}$ we have $\omega > 1$. Now we may develop ω into a regular continued fraction

$$\omega =: [a_n, a_{n+1}, \ldots]$$

with $a_n \ge 1$. By Proposition 3.1 it follows that

$$\xi_0 = [a_0, a_1, \ldots, a_{n-1}, a_n, \ldots], \tag{3.24}$$

hence $\frac{c}{d} = \frac{A_{n-1}}{B_{n-1}}$ is an approximating fraction for ξ_0. $\square$

3.4 The Algorithm

First we will show a trick that reduces the determination of the prime factors to the calculation of the period of a certain number theoretic function. Namely, assume x is coprime to n and define the exponential function (with base x) modulo n:

$$f_n(a) := x^a \pmod{.n}.$$

It is well-known that the sequence $\{f_n(a)\}_{a \in \mathbb{N}_0}$ is periodic. The length r of the period is called "the period of $x \pmod{.n}$". Assume r is even. Then (by Fermat's Little Theorem) we have that

$$(x^{r/2})^2 = 1(mod.n)$$

and hence

$$(x^{r/2} - 1)(x^{r/2} + 1) = 0(mod.n). \tag{3.25}$$

This means that (unless $x^{r/2} = \pm 1(mod.n)$), then at least one of the terms $x^{r/2} \pm 1$ must have a nontrivial factor in common with n. (Note that $x^{r/2} = 1$ would be the case if $r/2$ were already the period or a multiple of it, that is why in the algorithm it will be important to really find the genuine period and not a multiple of it.) So as soon as we have determined the period r, we have a good chance of finding a factor of n by computing (by the Euclidean algorithm, e.g.) the numbers $\gcd(x^{r/2} \pm 1, n)$. So our goal must be to determine efficiently the period of exponential functions modulo n (where n is the number to be factorized).

Now we are prepared for presenting Shor's factorization algorithm in detail. Let $n = pq$ (p, q prime) be the number to be factorized.

Shor's Factorization Algorithm

– Choose a number d with small prime factors such that $2n^2 \leq d \leq 3n^2$.
– Choose a random integer x that is coprime to n.
– Repeat the following steps $\log d$ times using the same x every time:
  – Create a quantum memory register of $2d$ non-negative integers modulo n and partition it into two halves called reg1 and reg2. For the state of the whole register we will write the ket vector $|reg1, reg2\rangle$.
  – Load reg1 with the integers $0, 1, \ldots, d-1$ and reg2 with zeroes at all places, afterwards normalize the register such that we may write (with a little abuse of notation) the state of the whole register as ket vector

$$|\psi\rangle = \frac{1}{\sqrt{d}} \sum_{a=0}^{d-1} |a, 0\rangle.$$

– Perform the transformation $x \mapsto x^a(mod.n)$ (using quantum parallelism) on each (non-normalized) number in reg1 and place the results to the corresponding places in reg2. Denote by r the period of the above transformation. Then the state of the (normalized) complete register becomes

$$|\psi\rangle = \frac{1}{\sqrt{d}} \sum_{a=0}^{d-1} |a, x^a(mod.n)\rangle.$$

– Measure the content of reg2 by the Hermitian operator A. Then this collapses to some k and has the effect of projecting out the state of reg1 to be a superposition of exactly those values of a for which $x^a = k(mod.n)$. Hence the state of the complete register is

$$|\psi\rangle = \frac{1}{\#M} \sum_{a' \in M} |a', k\rangle,$$

where $M := \{a' : x^{a'} = k(mod.n)\}$.

– Compute the discrete (fast) Fourier transform of the projected state in
reg1 and put this result back to reg1. This maps the projected state in
reg1 into a superposition

$$|\psi\rangle = \frac{1}{\#M} \sum_{a' \in M} \frac{1}{\sqrt{d}} \sum_{h=0}^{d-1} e^{2\pi i a' h/d} |h, k\rangle.$$

– Now the Fourier transform in reg1 is a periodic function peaked at multiples of the inverse period $1/r$. States corresponding to integer multiples
of $1/r$ and those close to them appear with greater probability ampli-
tudes than those that do not correspond to integer multiples of the inverse period. So in each step, we get a number h' such that $\frac{h'}{d}$ is near
to the multiple $\frac{\lambda}{r}$ of the inverse period of the exponential map for some
$\lambda \in \mathbb{N}$. In order to estimate λ, one can compute the continued fraction
expansion of $\frac{h'}{d}$ as long as the denominator is less that n and then retain
the closest such fraction as $\frac{\lambda}{r}$. If this is done sufficiently often, we have
enough samples λ_i that lead to a guess of the true λ and thus of r.
– Now that we know r, we can determine the factors of n (with high probability) as demonstated at the beginning of this section.

We remark that of course (with rather low probability) Shor's algorithm can
fail. Such counterexamples are easily constructed. But they represent rather
untypical cases.

Furthermore, instead of using the classical (fast) Fourier transform, there
are also quantum algorithms for the Fourier transform, which make Shor's
algorithm working still faster in practice, but not to such an extent that
the linear order of complexity is even ameliorated. Seifert (2001) suggests
an approach where, in contrast to the above-mentioned algorithm, he uses
simultaneous diophantine approximation to reduce considerably the number
of qubits necessary.

4 Physical Random-Number Generators

4.1 Generalities

The doctrine in cryptology is that the algorithm of encryption is known to
the adversary (Eve) and that the only thing that is kept secret is the key,
which normally is a bitsequence or a sequence of natural numbers or elements of a finite ring (e.g. a residue ring or a finite field). Mostly, such
key sequences are produced by an algorithmic generator (i.e., they are so-
called pseudo-random numbers), since these offer the following benefits: the
sequence of numbers can be reproduced for debugging and testing; no special
hardware is necessary; a large quantity of random numbers can be produced
in a short time. In Chapter 7, we will provide several tests of "randomness"
of such pseudo-random sequences. However, there is no practically implementable "universal" test of randomness: every test procedure just measures
a certain aspect of "non-regularity". If one wants to have genuine random
numbers, then they have to be produced by a physical device. A very drastic
drawback of classical pseudo-random generators has been pointed out in the
paper entitled "Random numbers fall mainly in the planes" by Marsaglia
(1968). Possible physical random sources are electronic noise produced by a
semiconducting diode (Richter (1993)) or the impulses of a Geiger counter
in connection with a radioactive source (Inoue et al. (1983)). In the latter
paper, the authors propose a hardware implementation of this device, the
radioactive source consisting of a PG-508 pulse generator. Another device
using a Geiger counter has been described in Nisley (1990), the RM-60 Micro
Roentgen Radiation Monitor from Aware Electronics. Finally, there is HOT
BITS (see Walker (1996)), a source of random bits available via the Internet,
which uses beta radiation from the decay of Krypton-85.
The output of such a generator (which in the latter case leads directly to a
Poisson (for the number of events) resp. exponential (for the inter-occurence
waiting times) distribution) has to be processed further in order to obtain
standard uniform random numbers (digits, or reals in $[0, 1]$). Since the parameters of the distribution of the data is not known exactly, only a small
amount of this information is used (usually the last digit), to be on the safe
side, and so the yield of this method is relatively small. However, physically
generated random numbers are expensive and can not be produced in too

D. Neuenschwander: Prob. and Stat. Methods in Cryptology, LNCS 3028, pp. 47-55, 2004.
© Springer-Verlag Berlin Heidelberg 2004

high quantities. For example, the HOT BITS hardware produces only about 240 bits per second.

Modifying an idea of von Neumann (1963), used to extract unbiased bits from a sequence of biased ones by comparison of two subsequent bits, we propose to obtain random numbers in $[0, 1]$ from a sequence $X_0, X_1, X_2, \ldots$ of independent exponentially distributed data by using $U_n := \frac{X_{2n}}{X_{2n}+X_{2n+1}}$. This gives us one real number for every two data values instead of only two bits, considerably increasing the output. If the distribution of the X_n is exponential, the U_n are uniform in $[0, 1]$. The question of the "rate of disappearing" of the bias (so-called extraction rate) is addressed in Section 4.3, in particular for rational biases b. It turns out that the size of b does not influence the extraction rate, but that the latter is solely determined by the arithmetic properties of b. On the other hand, the extraction rate can be shown to be 0 in Lebesgue-almost all cases.

In the practical implementation of this method we have to take into account that the exponential times X_n can only be measured up to a certain precision.

4.2 Construction of Uniformly Distributed Random Numbers from a Poisson Process

In this section, we will consider the output of a Geiger counter as source of randomness. The other examples mentioned in the previous section are of a similar nature. If the number of impulses during a fixed amount of time is counted, a variable with a Poisson distribution is the raw material that has to be processed further in order to obtain unbiased random bits. Usually, the length t_0 of the time interval is chosen large with respect to the mean time $1/\theta$ between two impulses; then the number N of counts during this interval has a Poisson distribution π_λ with $\lambda = t_0\theta$. In most cases, the last digit X in the binary representation of N is used as an approximation for a uniformly distributed random bit.

Another method (see Inoue et al. (1983)) makes use of the random waiting time T between two consecutive impulses, which obeys an exponential distribution ε_θ. Clearly, if the intensity $\theta > 0$ were known exactly, one could obtain a uniformly distributed random variable U just by the usual transform method $U := \exp(-\theta T)$. But θ not being known exactly enough to guarantee that U is "sufficiently" uniform, it has to be estimated. One can use two consecutive waiting times produced by the Geiger counter, one so to say to estimate θ and the other one to obtain a uniform random variable.

The following lemma is easy:

Lemma 4.1. *Let X and Y be independent random variables with common exponential distribution ε_θ (where $\theta > 0$). Then*

$$U := \frac{X}{X + Y}$$

obeys a uniform law on $[0, 1]$.

Therefore if the raw material is a stream of independent exponential random times X_n ($n \in I\!N$), a sequence of independent uniform variables can be obtained by setting $U_n := X_{2n}/(X_{2n} + X_{2n+1})$.

Unfortunately the waiting time between two impulses of the Geiger counter is not measured as a real number, but only in multiples of the length Δ of the clock cycle (w.l.o.g. we may assume $\Delta = 1$). If two impulses occur during one clock cycle, then they are counted as one. Hence the n-th observation of an impulse occurs at the time S'_n defined recursively by $S'_0 := 0$,

$$S'_n := \min\{k \in I\!N : N_k \ge N_{S'_{n-1}} + 1\},$$

where the Poisson process $\{N_t\}_{t \ge 0}$[1] indicates the number of impulses up to time t. Instead of the sequence $\{X_n\}_{n \ge 1}$ of exactly exponentially distributed waiting times between two impulses, we can only observe the sequence $\{X'_n\}_{n \ge 1}$, where $X'_n := S'_n - S'_{n-1}$.

Proposition 4.1. *The* $X'_1, X'_2, \ldots$ *are i.i.d. such that* $X'_n - 1$ *obeys a geometric distribution with parameter* $1 - \exp(-\theta)$.

Proof: Let $\{\mathcal{F}_t\}_{t \ge 0}$ denote the canonical filtration of the Poisson process $\{N_t\}_{t \ge 0}$. Then $S'_1 < S'_2 < \ldots$ is a sequence of stopping times. Assume $n \ge 2$. Since the Poisson process is stationary with independent increments, the process $\{N'_t\}_{t \ge 0}$ with $N'_t := N_{t + S'_{n-1}} - N_{S'_{n-1}}$ is again a Poisson process with parameter θ. Therefore the distribution of

$$X'_n = S'_n - S'_{n-1} = \min\{k \in I\!N : N'_k \ge 1\}$$

is the same as that of X'_1. The latter law can easily be calculated to be the geometric distribution with parameter

$$P(X'_1 - 1 = 0) = P(X_1 < 1) = 1 - \exp(-\theta).$$

---

[1] A stochastic process $\{N_t\}_{t \ge 0}$ is called a Poisson process with intensity $\lambda > 0$ if N_t obeys a Poisson distribution with parameter λt ($t > 0$). This is equivalent to the fact that for the "jump times"

$$\Gamma_0 = 0 < \Gamma_1 < \Gamma_2 < \ldots$$

(where $\Gamma_k := \inf\{t \ge 0 : N_t \ge k\}$) we have that the "inter-occurence times" $\Gamma_{k+1} - \Gamma_k$ are i.i.d exponentially distributed random variables as

$$P(\Gamma_{k+1} - \Gamma_k > x) = e^{-\lambda x}$$

for $x \ge 0$.

Furthermore, the process $\{N'_t\}_{t\geq0}$ and hence the random variable X'_n is independent of $\mathcal{F}_{S'_{n-1}}$. On the other hand, the random variables $X'_1, X'_2, \ldots, X'_{n-1}$ are $\mathcal{F}_{S'_{n-1}}$-measurable and hence independent of X'_n. $\square$

Denote by F_X the distribution function of a random variable X.

Theorem 4.1. *Let $X' - 1, Y' - 1$ be independent geometric random variables with parameter $\theta' = 1 - \exp(-\theta)$ and denote by U a random variable distributed uniformly on the interval $[0, 1]$. Then $U' := \frac{X'-0.5}{X'+Y'-1}$ satisfies*

$$\frac{1}{2}\tanh\left(\frac{\theta}{2}\right) \leq ||F_{U'} - F_U||_\infty \leq 1 - \exp\left(-\frac{\theta}{2}\right).$$

Proof: The lower bound follows from the observation that F_U is continuous at $\frac{1}{2}$ whereas $F_{U'}$ has a jump of size $P\{U' = \frac{1}{2}\} = \sum_n P\{X' = n\}P\{Y' = n\} = \sum_{n=0}^\infty \theta'^2(1 - \theta')^{2n} = \frac{\theta'}{2-\theta'} = \frac{1-\exp(-\theta)}{1+\exp(-\theta)} = \tanh\left(\frac{\theta}{2}\right)$.

We will now assume w.l.o.g. that X' and Y' are of the form $X' = \lfloor X \rfloor$ and $Y' = \lfloor Y \rfloor$ with X, Y independent and with exponential distribution with parameter θ. Let $U := X/(X+Y)$. Since the distribution of U' is symmetric about $\frac{1}{2}$ it is easy to see that for the upper bound it is sufficient to show $|F_{U'}(t) - t| \leq 1 - \exp(-\theta/2)$ only for $t \in]0, \frac{1}{2}]$. For such t we have

$$F_{U'}(t) - t = E\left(1_{[0,t]}(U') - 1_{[0,t]}(U)\right)$$

$$= \sum_{m,n\in\mathbb{N}} E\Big([1_{[0,t]}\big(\frac{m+0.5}{m+n+1}\big) - 1_{[0,t]}(U)]$$

$$\cdot 1_{\{m<X<m+1,\ n<Y<n+1\}}\Big)$$

$$= S_+ - S_-,$$

where

$$S_+ = \sum_{\frac{m+0.5}{m+n+1}\leq t,\ \frac{m+1}{m+n+1}>t} P(m < X < m+1,\ n < Y < n+1,\ \frac{X}{X+Y} > t)$$

and

$$S_- = \sum_{\frac{m+0.5}{m+n+1}>t,\ \frac{m}{m+n+1}<t} P(m < X < m+1,\ n < Y < n+1,\ \frac{X}{X+Y} \leq t).$$

The last equality follows from the fact that the random variable in the expectation takes only the values $-1, 0,$ or 1, and all summands with either $\frac{m+1}{m+n+1} \leq t$ or $\frac{m}{m+n+1} \geq t$ vanish, since the square $]m, m + 1[\times]n, n + 1[$ lies completely on one side of the line $\{\frac{x}{x+y} = t\}$ in these cases.

We now collect the summands in $S_\pm$ that belong to the same m. Let $a := (1 - t)/t \geq 1$; then $\frac{x}{x+y} > t$ if and only if $ax > y$.

From this we obtain

$$S_+ = \sum_{m=0}^{\infty} P(X < m+1, \; Y > n_m, \; aX > Y)$$

and

$$S_- = \sum_{m=0}^{\infty} P(X > m, \; Y < n_m, \; aX \leq Y),$$

where n_m is the smallest $n \in I\!N$ with $\frac{m+0.5}{m+n+1} \leq t$. Considering a single summand we have

$$
\begin{aligned}
&P(X > m, \; Y < n_m, \; aX \leq Y) \\
&\leq P(m < X < m + (1/2), \; am < Y < n_m) \\
&= c \cdot P(m < X < m+1, \; am < Y < n_m)
\end{aligned}
$$

with

$$c := P(m < X < m + (1/2))/P(m < X < m + 1) = \left(1 + \exp\left(-\frac{\theta}{2}\right)\right)^{-1}.$$

In order to prove the latter inequality we have used the fact that the density of P is a decreasing function of $x + y$ and an elementary geometric argument in the $m - n_m$-plane. A similar argument yields

$$P(X < m+1, \; Y > n_m, \; aX > Y) \leq c \cdot P(m < X < m+1, \; n_m < Y < a(m+1)).$$

Summing up, we obtain

$$
\begin{aligned}
S_+ + S_- &\leq \sum_{m=0}^{\infty} c \cdot P(m < X < m+1, \; am < Y < n_m) \\
&\qquad\qquad + c \cdot P(m < X < m+1, \; n_m < Y < a(m+1)) \\
&= c \cdot \sum_{m=0}^{\infty} P(m < X < m+1, \; am < Y < a(m+1)) \\
&= c \cdot \sum_{m=0}^{\infty} \exp(-\theta(a+1)m)(1 - \exp(-\theta))(1 - \exp(-a\theta)) \\
&= \frac{(1 - \exp(-\theta))(1 - \exp(-a\theta))}{(1 + \exp(-\theta/2))(1 - \exp(-(a+1)\theta))}.
\end{aligned}
$$

But since $1 - \exp(-a\theta) \leq 1 - \exp(-(a+1)\theta))$, we finally get the bound

$$|F_{U'}(t) - t| = |S_+ - S_-| \leq S_+ + S_- \leq 1 - \exp\left(-\frac{\theta}{2}\right),$$

and this proves Theorem 4.1. $\square$

The upper and lower bounds $1 - \exp\frac{-\theta}{2} \leq \frac{\theta}{2}$ and $\frac{1}{2}\tanh\frac{\theta}{2} \approx \frac{\theta}{4}$ in Theorem 4.1 differ by a factor of approximately 2. As one can see from numerical experiments, the lower of these is the true value, but the proof of this fact is more complicated.

4.3 *The Extraction Rate for Biased Random Bits

In this section, it will make things a little simpler (e.g., as we will see, we can work with expectations) if we replace $\mathbb{B} = \{0, 1\}$ by $\mathcal{B} := \{1, -1\}$. Since there will be no danger of misunderstanding, also the elements of $\mathcal{B}$ will be called (random) "bits".

We want to investigate the following question: Given n i.i.d. random bits with common bias b (i.e. $P(X_1 = 1) - P(X_1 = -1) = E(X_1) = b \in]0, 1[$ (w.l.o.g.)), how is it possible to construct from them an "as unbiased as possible" random bit? It turns out that a good method is to multiply[2] the $X_i \in \mathcal{B}$, for if

$$P_n := \prod_{i=1}^{n} X_i,$$

then the bias of P_n turns out to be only b^n, i.e. $P(P_n = 1) - P(P_n = -1) = b^n$. One may ask if there are functions $f : \mathcal{B}^n \to \mathcal{B}$ that behave better (in the sense of bias reduction) than multiplication. Let us define, for f and b as defined before, the quantity

$$\xi_{f,n}(b) := |E(f(X_1, X_2, \ldots, X_n))|$$

and

$$\Xi_n(b) := \min_{f:\mathbb{B}^n \to \mathbb{B}} \xi_{f,n}(b).$$

The relation $\xi_{\cdot,n}(b) = b^n$ (as mentioned before) can be interpreted as follows: For each new (independent b-biased) bit source X_{n+1} combined with the sources $X_1, X_2, \ldots, X_n$, the multiplication-function "extracts" another factor b in the output bit P_{n+1} (compared with P_n). So if we replace the multiplication-function by an (asymptotically (as $n \to \infty$)) optimal function f, we should have at least the extra multiplicative factor b for every step $n \to n+1$ (i.e. by taking one additional bit source). Therefore, we define the so-called extraction rate of b by

$$\Xi(b) := \lim_{n \to \infty} \sqrt[n]{\Xi_n(b)}.$$

The extraction rate can be interpreted as the optimal asymptotic multiplicative effect of each new input bit source on the resulting bias of the output bit. Or - in other words - it is the asymptotical (as $n \to \infty$) speed of the diminution of the bias per new random bit source, when the final output bit is produced by adding (mod.2) (in $\mathbb{B}$) n independent biased random bit sources. It can be shown that for Lebesgue-almost all $b \in]0, 1[$ we have $\Xi(b) = 0$ (see Näslund, Russell (2001), Theorem 21). For rational b we have the following:

---

[2] If we identify $\mathcal{B}$ and $\mathbb{B}$ in the natural way, then multiplication in $\mathcal{B}$ corresponds to addition mod.2 in $\mathbb{B}$.

Theorem 4.2. *If* $b \in \mathbb{Q}$, $b = \frac{r}{s}$, $r, s \in \mathbb{N}$, r, s *relatively prime, then* $\Xi(b) = \frac{1}{s}$.

So interestingly enough, it is not the size of b, but rather its arithmetic properties that determine its extraction rate!

Proof of Theorem 4.2: 1. We first prove that

$$\Xi_n(b) \geq \frac{1}{s^n}. \tag{4.1}$$

Let us fix some notation. For a subset $C \subset \mathcal{B}^n$, define its weight by

$$w(C) := P((X_1, X_2, \ldots, X_n) \in C),$$

and put

$$f(X_1, X_2, \ldots, X_n) := 2(1(C)(X_1, X_2, \ldots, X_n) - \frac{1}{2}).$$

Now consider a collection (subset) $C \subset \mathcal{B}^n$ with $w(C) = \frac{1+\delta}{2}$, where $|\delta|$ is the bias of f. W.l.o.g. we may suppose that $(-1, -1, \ldots, -1) \notin C$. Then we may calculate

$$w(C) = \frac{1 + \delta}{2}$$

$$= \sum_{i=1}^{n} t_i (\frac{r}{s})^i (1 - \frac{r}{s})^{n-i}$$

$$= \frac{1}{s^n} \sum_{i=1}^{n} t_i r^i (s - r)^{n-i}$$

for some integers $t_i \in \{0, 1, \ldots, \binom{n}{i}\}$, or - equivalently -

$$s^n (1 + \delta) = 2 \sum_{i=1}^{n} t_i r^i (s - r)^{n-i}.$$

Since $\delta \neq 0$ and $b > 0$ we have that $r > 1$ is a divisor of the right-hand side of the above equality. Furthermore, we have supposed that r and s are relatively prime. So the left-hand side must be an integer (since the right-hand side is) and inequality (4.1) follows.

2. Now we turn to the other direction. We will construct a family of functions $f_n : \mathcal{B}^n \to \mathcal{B}$ with the property that

$$\sqrt[n]{|E(f_n(X_1, X_2, \ldots, X_n))|} \to \frac{1}{s}. \tag{4.2}$$

For this, we will prove the following lemma, which is also of some independent interest. Then (4.1) and (4.2) will yield the result of Theorem 4.2. □

Lemma 4.2. *If b is as in Theorem 4.2, we have that*

$$\Xi(b) \leq \frac{1}{s}.$$

More precisely, for $n > 2r + 1$ we obtain

$$\Xi_n(b) \leq \frac{2r(s-r)^2}{s^n}$$

and there exists a (deterministic) polynomial-time algorithm for finding an optimal f, such that

$$\xi_{f,n}(b) \leq \frac{2r(s-r)^2}{s^n}.$$

Proof: Define $q := s - r$, so that we have $\frac{q}{s} + \frac{r}{s} = 1$. Since we have supposed $b > 0$, it follows that $r > q$. Let $\mathcal{B}_i^{(n)}$ be the i-th level of $\mathcal{B}^n$, i.e. those elements of $\mathcal{B}^n$ with Hamming weigth (number of ones) i. Let $P_i^{(n)}(b)$ denote the probability that an element of $\mathcal{B}^n$ is equal to some fixed element of $x \in \mathcal{B}_i^{(n)}$. This probability is indeed independent of the specific x and given by

$$P_i^{(n)}(b) = b^i (1-b)^{n-i}.$$

Hence in our case, we have

$$P_i^{(n)}(b) = \frac{r^i q^{n-i}}{s^n}.$$

We want to find collections $C_n \subset \mathcal{B}^n$ such that $s^n w(C_n)$ is "close" to $\frac{s^n}{2}$. Then for the function

$$f_n(X_1, X_2, \ldots, X_n) := 2(1(C_n)(X_1, X_2, \ldots, X_n) - \frac{1}{2})$$

we will have that $E(f_n(X_1, X_2, \ldots, X_n))$ will be close to 0.
For this construction we proceed as follows: Define an initial collection

$$\tilde{C}_n := \mathcal{B}_n^{(n)} \cup \mathcal{B}_{n-1}^{(n)} \cup \mathcal{B}_{n-2}^{(n)} \cup T,$$

where T is a maximal subset of $\bigcup_{i<n-2} \mathcal{B}_i^{(n)}$ for which $|\mathcal{B}_j^{(n)} \setminus T| \geq r - 1$ (for $1 \leq j \leq n - 3$) and $s^n w(\tilde{C}_n) \leq \frac{s^n}{2}$. Now let us adjust this collection suitably to bring its weight (multiplied by s^n) closer to $\frac{s^n}{2}$. Since $r > q$ and $P_i^{(n)}(b) < P_j^{(n)}(b)$ (for $i < j$) and by the maximality of T we get

$$|s^n w(\tilde{C}_n) - \lfloor \frac{s^n}{2} \rfloor| < s^n P_{n-2}^{(n)}(b) = r^{n-2} q^2.$$

Now consider the cyclic group $\mathbb{Z}_{r^{n-2}q^2}$ and denote by $\pi : \mathbb{Z} \to \mathbb{Z}_{r^{n-2}q^2}$ the canonical projection. Since r and q are relatively prime, it follows that for every $i > 2$, the element $\pi(r^{n-i}q^i)$ has order r^{i-2} in $\mathbb{Z}_{r^{n-2}q^2}$, so that we obtain the following chain (or "tower") of subgroups of $\mathbb{Z}_{r^{n-2}q^2}$:

$$0 = \langle \pi(r^{n-2}q^2) \rangle \subset \langle \pi(r^{n-3}q^3) \rangle \subset \ldots \subset \langle \pi(rq^{n-1}) \rangle$$

($\langle x \rangle$ denotes the cyclic subgroup generated by x). All the groups in the above chain have index r in the next one and the last one ($\langle \pi(rq^{n-1}) \rangle$) has index rq^2 in $\mathbb{Z}_{r^{n-2}q^2}$. Thus the group $\langle \pi(rq^{n-1}) \rangle$ can be used to approximate every element of $\mathbb{Z}_{r^{n-2}q^2}$ to within an additive error of rq^2. In particular for $\Delta :=$ $\pi(\lfloor \frac{s^n}{2} \rfloor - s^n w(\tilde{C}_n))$, there exists an element $\Delta' \in \langle \pi(rq^{n-1}) \rangle$ such that

$$\Delta - \Delta' \in \{\pi(0), \pi(1), \ldots, \pi(rq^2 - 1)\}.$$

On the other hand, we of course may write $\Delta' =: c\pi(rq^{n-1})$, so by well-known algebraic facts (see Näslund, Russell (2001), p. 308) one has an equation

$$\Delta' = \sum_{i=1}^{n-3} t_i \pi(r^i q^{n-i}) \in \langle \pi(rq^{n-1}) \rangle$$

with integers $t_i \in \{0, 1, \ldots, r-1\}$. As $r > q$, we may "lift" this equation to

$$\lfloor \frac{s^n}{2} \rfloor = \sum_{i=1}^{n-3} t_i r^i q^{n-i} - mr^{n-2}q^2 + w(\tilde{C}_n)s^n + E$$

(where $m \leq nr$ and $E \in \{0, 1, \ldots, rq^2 - 1\}$ represents the error term). Now, if we add t_i elements of $\mathcal{B}_i^{(n)}$ to $\tilde{C}_n$ and, on the other hand, remove m elements of $\mathcal{B}_{n-2}^{(n)}$ from $\tilde{C}_n$ (which is possible as long as $m < \binom{n}{n-2}$, i.e. $r < \frac{n-1}{2}$), we indeed obtain a new collection C_n with

$$s^n w(C_n) - \lfloor \frac{s^n}{2} \rfloor < E.$$

Dividing this equation by s^n yields

$$\Xi_n(b) \leq \frac{2rq^2}{s^n}$$

and the result follows (since each step of the above-described algorithm can be carried out in polynomial time). $\square$

5 Pseudo-random Number Generators

5.1 Linear Feedback Shift Registers

In contrast to Chapter 4, where we discussed genuine physical random number generators, here we will deal with so-called pseudo-random number generators. These are determinsitic algorithms that produce an output which behaves "more or less" like random numbers. The advantage is that like that, much more data can be generated per time unit than with physical devices. On the other hand, pseudo-random numbers never have the quality of genuine random numbers. There is a definition of "perfect pseudo-randomness" in the sense that, loosely speaking, a source is perfectly pseudo-random if it can not "efficiently" be distinguished from genuine random numbers. We will deal with that in more detail in Section 5.3. However, this test is not practically implementable. In reality, one can only test for finitely many necessary conditions for a source to be considered as "sufficiently random". Normally, one tries to generate a uniform random variable U on the interval $[0, 1[$ (more precisely: one approximates U by a discrete uniform distribution on the set $\{\frac{k}{N} : 0 \leq k \leq N - 1\}$, where N is chosen sufficiently large).

In the sequel, w.l.o.g. we will interpret (pseudo-)random numbers as bitsequences $\{x_i\}_{i\geq 0} = \{x_0 x_1 \ldots\}$ ($x_i \in I\!B = \{0, 1\}$).

A catalog of some "minimal" requests for pseudo-random generators was established 1967 by S. Golomb. For this, we need some preparation. As a computer has only finitely many memory places, every pseudo-random sequence generated by a computer is eventually periodic. In the sequel, we consider only periodic pseudo-random number generators. Let p be the period of the pseudo-random number generator, i.e., the smallest natural number with the property $x_{i+p} = x_i$ ($\forall i \geq 0$). A run is here, by definition, a sequence of identical elements of $I\!B$; we will speak of a block, resp. a gap, if it is a run of ones, resp. zeroes. Let $A(k)$ resp. $D(k)$ be the number of matches, resp. non-matches, of $\{x_i\}_{i\geq 0}$ and $\{x_{i+k}\}_{i\geq 0}$ counted over a whole period:

$$A(k) := |\{i \in \{0, 1, \ldots, p - 1\} : x_i = x_{i+k}\}|,$$

$$S(k) := |\{i \in \{0, 1, \ldots, p - 1\} : x_i \neq x_{i+k}\}|.$$

D. Neuenschwander: Prob. and Stat. Methods in Cryptology, LNCS 3028, pp. 57-75, 2004.
© Springer-Verlag Berlin Heidelberg 2004

The autocorrelation $AC(k)$ of the periodic sequence $\{x_i\}_{i \geq 0}$ is defined as

$$AC(k) = \frac{A(k) - D(k)}{p}.$$

If k is a multiple of the period length p, then one speaks of "in-phase"-autorcorrelation; in this case we always have $AC(k) = 1$. In the other case, $AC(k)$ is called "out-of-phase"-autocorrelation; it lies always between -1 and 1. Now Golomb's conditions are the following:

(G1) The number of zeroes and the number of ones per period are $p/2$ (if p is even) and $(p \pm 1)/2$ (if p is odd) (i.e. zeroes and ones appear with approximately the same probability).

(G2) In a cycle, half of the runs have length 1, a quarter of them length 2, an eighth length 3, ... Half of the runs of a certain length are blocks, the other half gaps. (This condition says that, e.g., after 01, the zero again has the same probability as the one, etc.)

(G3) The out-of-phase-autocorrelation $AC(k)$ has the same value for all k. (This can be interpreted as follows: If one counts the number of matches between a sequence and its shift by k places, one does not obtain any information about the period p of this sequence (with the exception if k is a multiple of p).)

We now consider bitsequences that are generated by so-called linear feedback shift registers (LFSR). The advantage of LFSR is that they are very easily implementable in hardware and work very fastly. An LFSR of length n has a vector of n places of memory. At the beginning, the initial state vector $(x_0, x_1, \ldots, x_{n-1}) \in \mathbb{B}^n$ is stored. The most important part of an LFSR is the so-called (linear) feedback function:

$$f : \mathbb{B}^n \ni (x_0, x_1, \ldots, x_{n-1}) \mapsto f(x_0, x_1, \ldots, x_{n-1}) = \sum_{i=0}^{n-1} c_i x_i \in \mathbb{B},$$

where $c_0, c_1, \ldots, c_{n-1} \in \mathbb{B}$ are fixed (built-in) parameters. After the first step, the LFSR will give the leftmost bit x_0 as output, delete this bit from its memory place, shift the contents of all the other memory places one place to the left, and put the value $x_n = f(x_0, x_1, \ldots, x_{n-1})$ in the rightmost memory place. (We note that instead of bits, one can work with elements of an arbitrary finite ring. Then the classical linear congruence generators are just shift registers of length $n = 1$ over a residue ring.) W.l.o.g. we may assume $c_0 = 1$, for otherwise one could replace the LFSR by a LFSR of length $n - 1$. An output sequence of an LFSR will be called a pseudo-noise sequence (PN-sequence) if it has the (maximal possible) period $p = 2^n - 1$. Since a LFSR generating a PN-sequence must assume all its possible states and since the output sequence is uniquely determined by the initial state, PN-sequences are automatically periodic. In the sequel, we want to investigate which LFSR generate PN-sequences. For this, we introduce the important notion of the characteristic polynomial (or recursion polynomial) of a LFSR: If one puts $c_n := 1$, then the polynomial

$$f(z) := \sum_{i=0}^{n} c_i z^i$$

is called the characteristic polynomial (or recursion polynomial) of the LFSR. With $\Theta(f)$ we denote the set of all possible output sequences of the LFSR with characteristic polynomial f:

$$\Theta(f) = \{\{x_i\}_{i \geq 0} : x_{k+n} = \sum_{i=0}^{n-1} c_i x_{k+i} \ (k \geq 0)\}.$$

One sees easily that $\Theta(f)$ is a vector space of dimension n over the field $\mathbb{B}$. If $f(z) = \sum_{i=0}^{n} c_i z^i$ denotes a polynomial with coefficients in $\mathbb{B}$, then we define the corresponding dual polynomial $f^*(z)$ by

$$f^*(z) = z^n f(1/z) = \sum_{i=0}^{n} c_i z^{n-i}.$$

Clearly, $f^{**}(z) = f(z)$ and $(f \cdot g)^*(z) = f^*(z) \cdot g^*(z)$. For the output sequence $\{x_i\}_{i \geq 0}$ we consider the generating function

$$S(z) = \sum_{i=0}^{\infty} x_i z^i$$

(interpreted as formal power series).

Lemma 5.1. *If one puts*

$$\tau(z) := \sum_{j=0}^{n-1} (\sum_{\ell=0}^{j} c_{n-\ell} x_{j-\ell}) z^j,$$

then it follows that

$$S(z) = \frac{\tau(z)}{f^*(z)}.$$

Proof:

$$S(z) f^*(z) = (\sum_{k=0}^{\infty} x_k z^k)(\sum_{\ell=0}^{n} c_{n-\ell} z^\ell)$$

$$= \sum_{j=0}^{\infty} (\sum_{\ell=0}^{\min\{j,n\}} c_{n-\ell} x_{j-\ell}) z^j$$

$$= \sum_{j=0}^{n-1} \sum_{\ell=0}^{j} c_{n-\ell} x_{j-\ell} z^j + \sum_{j \geq n} \sum_{\ell=0}^{n} c_{n-\ell} x_{j-\ell} z^j$$

$$= \tau(z) + \sum_{j \geq n} (\sum_{i=0}^{n} c_j x_{(j-n)+i}) z^i$$

$$= \tau(z). \square$$

As $|\Theta(f)| = 2^n$ and since there are exactly 2^n polynomials of degree $< n$ over $\mathbb{B}$, one obtains (by identifying the output sequence with its generating function) the following:

Corollary 5.1.

$$\Theta(f) = \{\frac{\tau(z)}{f^*(z)} : \deg \tau(z) < n\}.$$

Lemma 5.2. *Suppose* $\{x_i\}_{i \geq 0} \in \Theta(f)$, $\{y_i\}_{i \geq 0} \in \Theta(g)$. *Then*

$$\{x_i + y_i\}_{i \geq 0} \in \Theta(\mathrm{lcm}(f,g)).$$

Proof: Based on Corollary 5.1, we write $S(z) = \alpha(z)/f^*(z)$, $T(z) = \beta(z)/g^*(z)$ (where $\deg \alpha(z) < \deg f(z)$, $\deg \beta(z) < \deg g(z)$). Furthermore, we put $h = \mathrm{lcm}(f,g)$ and define the polynomials u and v by $h = u \cdot f$ and $h = v \cdot g$. As

$$S(z) + T(z) = \frac{\alpha(z)}{f^*(z)} + \frac{\beta(z)}{g^*(z)}$$
$$= (\alpha(z)u^*(z) + \beta(z)v^*(z))/h^*(z),$$

and, on the other hand, $\alpha(z)u^*(z)$ and $\beta(z)v^*(z)$ both have lower degree than $h(z)$, it follows that $S(z) + T(z) \in \Theta(h)$. $\square$

From the theory of finite fields, the following is known:

Lemma 5.3. *(i) For every polynomial* $f(z)$ *with* $f(0) = 1$ *there exists an* $m \in \mathbb{N}$ *such that* $f(z)$ *is a divisor of* $z^m + 1$. *The smallest such* m *is called the period of the polynomial* $f(z)$.
(ii) If $n := \deg f(z)$ *and if* $f(z)$ *is irreducible, then the period of* $f(z)$ *divides* $2^n - 1$. *If the irreducible polynomial* $f(z)$ *has the (maximal) period* $2^n - 1$, *then the polynomial* $f(z)$ *is called primitive.*
(iii) The number of primitive irreducible polynomials of degree n *is given by* $\varphi(2^n - 1)/n$ *(φ denoting the Euler totient function).*

Lemma 5.4. *If the polynomial* $f(z)$ *has period* m *and degree* n *and if* $\{x_i\}_{i \geq 0} \in \Theta(f)$, *then the period of* $\{x_i\}_{i \geq 0}$ *divides* m.

Proof: Let $g(z)$ be a polynomial such that

$$z^m + 1 = f(z) \cdot g(z) \tag{5.1}$$

and with degree $m - n$. If on both sides of (5.1) one passes to the dual polynomial, one obtains

$$z^m + 1 = f^*(z) \cdot g^*(z).$$

By Lemma 5.1 there is a polynomial $\tau(z)$ of degree $< n$ such that

$$
\begin{aligned}
S(z) &= \frac{\tau(z)}{f^*(z)} \\
&= \frac{\tau(z) \cdot g^*(z)}{1 + z^m} \\
&= \tau(z) \cdot g^*(z) \cdot (1 + z^m + z^{2m} + \ldots).
\end{aligned}
$$

Since $\deg g^*(z) = m - n$, it follows that $\deg(\tau(z) \cdot g^*(z)) < m$. So the period of $S(z)$ divides m. $\square$

Lemma 5.5. *If the irreducible polynomial $f(z)$ has period m and degree n and if $\{x_i\}_{i \geq 0} \in \Theta(f)$, then $\{x_i\}_{i \geq 0}$ has period m.*

Proof: Let p denote the period of $\{x_i\}_{i \geq 0}$. By Lemma 5.4, p divides m. Hence

$$S(z) = \frac{u(z)}{1 + z^p} \tag{5.2}$$

for a suitable polynomial $u(z)$ of degree $< p$. One the other hand, due to Lemma 5.1 we have

$$S(z) = \frac{\tau(z)}{f^*(z)}. \tag{5.3}$$

Comparing (5.2) and (5.3) yields

$$(1 + z^p) \cdot \tau(z) = u(z) \cdot f^*(z)$$

and hence (by passing to the dual polynomial on both sides)

$$(1 + z^p) \cdot \tau^*(z) = u^*(z) \cdot f(z).$$

As $f(z)$ is irreducible and $\tau^*(z)$ has degree $< n$, it follows that $f(z)$ is a divisor of $z^p + 1$. Since $f(z)$ has period m, we obtain that m divides p. But as p is also a divisor of m (as seen before), the assertion follows. $\square$

Lemma 5.6. *If $f(z)$ is a polynomial of degree n and if $\{x_i\}_{i \geq 0} \in \Theta(f)$ is a PN-sequence, then $f(z)$ is irreducible.*

Proof: From the theory of factorization in rings (here applied to rings of polynomials), it follows that there exists an irreducible polynomial $f_1(z)$ with positive degree n_1 and a polynomial $f_2(z)$ such that $f(z) = f_1(z) \cdot f_2(z)$. By Corollary 5.1 we have that $1/f_1^*(z) \in \Theta(f_1)$, so by Lemmas 5.3(ii) and 5.4, the period of $1/f_1^*(z)$ divides $2^{n_1} - 1$. On the other hand $1/f_1^*(z) = f_2^*(z)/f^*(z) \in \Theta(f)$, so $1/f_1^*(z)$ must be a shift of $\{x_i\}_{i \geq 0}$ and thus have period $2^n - 1$. It follows that $n = n_1$, thus $f(z) = f_1(z)$. $\square$

So by Lemmas 5.5 and 5.6, we obtain the following theorem:

Theorem 5.1. *The output sequence of a LFSR is a PN-sequence iff the characteristic polynomial is primitive.*

Due to Lemma 5.3(iii), there are thus exactly $\varphi(2^n - 1)/n$ different LFSR of length n that generate PN-sequences.

One can show that LFSR that generate PN-sequences satisfy Golomb's conditions (G1)-(G3):

(G1): Since every state occurs exactly once per period and since the leftmost bit always yields the next output bit, it follows that the number of ones, resp. zeroes, per period is 2^{n-1}, resp. $2^{n-1} - 1$.

(G2): There are 2^{n-k-2} states whose leftmost $k + 2$ bits have the form $011\ldots10$, resp. $100\ldots01$. So gaps and blocks of length $k \leq n - 2$ occur exactly 2^{n-k-2} times per period. The state $011\ldots1$ occurs exactly once. Its successor state is $11\ldots1$, after which the state $11\ldots10$ follows. Hence there is no block of length $n - 1$ and 1 block of length n. By analogy, there exists 1 gap of length $n - 1$ and no gap of length n.

(G3): If $\{x_i\}_{i\geq0} \in \Theta(f)$, then also $\{x_{i+k}\}_{i\geq0} \in \Theta(f)$ and thus (since $\Theta(f)$ is a vector space) $\{x_i + x_{i+k}\}_{i\geq0} \in \Theta(f)$. The number of matches per period between $\{x_i\}_{i\geq0}$ and $\{x_{i+k}\}_{i\geq0}$ is equal to the number of zeroes of $\{x_i + x_{i+k}\}_{i\geq0}$ per period, which by (G1) has the value $2^{n-1} - 1$. By analogy, the number of non-matches is 2^{n-1}. So the out-of-phase-autocorrelation assumes the value

$$AC(k) = -\frac{1}{2^n - 1} \quad (1 \leq k < 2^n - 1).$$

Of course, every finite bitsequence can be produced by an LFSR. The length of the shortest such LFSR can be determined by the Berlekamp-Massey algorithm (see Section 7.11) and is called the linear complexity of the bitsequence. Non-linear filtering of PN-sequences can lead to high linear complexity (see Kalouptsidis, Kolokotronis (2003)).

5.2 The Shrinking and Self-shrinking Generators

The shrinking generator consists of two LFSR over $GF(2)$, an LFSR $a = (a(0), a(1), \ldots)$ and a second one (called the selector) $s = (s(0), s(1), \ldots)$. Now the output of the generator will be the x-sequence, which is a "shrunken" version of the a-sequence, in the sense that the element $a(i)$ will be included in the x-sequence if $s(i) = 1$, otherwise it will be discarded. In other (more formal) words:

$$x(k) := a(i_k),$$

where i_k denotes the position of the k-th 1 in the selector sequence s. The shrinking generator is easy to implement and has, as we will see, good statistical properties. First, we will investigate the period and the linear complexity of the x-sequence. Let T_a, resp. $|a|$, denote the period, resp. length of the LFSR a (and analogously for s and x).

Theorem 5.2. *If a and s have primitive characteristic polynomials and if T_a and T_s are relatively prime, then*

$$T_x = (2^{|a|} - 1)2^{|s|-1}.$$

Proof: W.l.o.g. we may assume that

$$|a| > \log_2 |s|. \tag{5.4}$$

Since the s-sequence has (due to the primitivity of its characteristic polynomial) $2^{|s|-1}$ elements 1 in a full period, one observes that

$$x(i + j2^{|s|-1}) = a(k_i + jT_s). \tag{5.5}$$

Furthermore, if for any indexes k, k' we have that $a(k + jT_s) = a(k' + jT_s)$ for all j, then it follows that

$$T_a | k - k'. \tag{5.6}$$

[Since the characteristic polynomial of a is primitive and since T_a and T_s are relatively prime, it follows that the characteristic polynomial of the sequence $\{a(k + jT_s)\}_{j \geq 0}$ is also primitive, hence this sequence also has period T_a.] Clearly, we have

$$T_x | T_a 2^{|s|-1}.$$

Since $x(i + j2^{|s|-1}) = x(i + T_z + j2^{|s|-1})$ for all i and j, together with (5.5) and (5.6) we obtain that

$$T_a | k_{i+T_x} - k_i \tag{5.7}$$

for all i. Or - in other words - for every i there exists a j_i such that

$$k_{i+T_x} = k_i + j_i T_a. \tag{5.8}$$

Replacing i by $i + 1$ in (5.8) yields

$$k_{i+1+T_x} = k_{i+1} + j_{i+1} T_a. \tag{5.9}$$

Now we subtract (5.8) from (5.9), giving

$$k_{i+T_x+1} - k_{i+T_x} = k_{i+1} - k_i + (j_{i+1} - j_i)T_a \tag{5.10}$$

for all i. On the one hand k_{i+T_x} and k_{i+T_x+1}, but on the other hand, k_i and k_{i+1} are also positions of consecutive ones in the s-sequence. So if $j_{i+1} - j_i \neq 0$, we would have at least T_a consecutive zeros somewhere in the s-sequence, which by assumption (5.4) has been ruled out. So $j_{i+1} = j_i$ and hence

$$k_{i+T_x+1} - k_{i+T_x} = k_{i+1} - k_i \tag{5.11}$$

for all i, which yields that the subsequences of s starting at the elements $s(k_i)$, resp. $s(k_i + T_x)$, are identical. This is only possible if $T_s | k_{i+T_x} - k_i$, hence the number of elements in the s-sequence between $s(k_i)$ and $s(k_i + T_x)$ is a multiple of the period $2^{|s|-1}$ of s. However, then the number of ones in this segment is a multiple of $2^{|s|-1}$. But on the other hand, this number is also T_x, so there exists a $t \in \mathbb{N}$ such that

$$T_x = t 2^{|s|-1}. \tag{5.12}$$

Relation (5.5) implies

$$a(k_0) = x(0) = x(jT_x) = x(jt2^{|s|-1}) = a(k_0 + jtT_s) \tag{5.13}$$

for all j. Thus $T_a | tT_s$ and hence (since T_a and T_s are supposed to be relatively prime) $T_a | t$, which, by (5.12), entails that $T_a 2^{|s|-1} | T_x$. $\square$

For the linear complexity L of the x-sequence, we get the following estimate:

Theorem 5.3. *Under the hypotheses of Theorem 5.2, we have*

$$|a| 2^{|s|-2} < L < |a| 2^{|s|-1}.$$

Proof: 1. Upper bound for L: In order to find an upper bound for L, we want to look for a polynomial $p(.)$ such that (by a little abuse of notation) $p(z) = 0$ for all possible outcomes of the sequence x (i.e., the coefficients of $p(z) = \sum_{k=0}^{n} c_i z^i$ represent the linear relation $\sum_{k=0}^{n} c_i x_i$ satisfied by the elements of the x-sequence). Let $x_{[s]}$ denote the sequence $\{x(j2^{|s|-1})\}_{j \geq 0}$. From (5.5), the elements of this sequence are all of the form $a(i + jT_s)$. By the hypothesis that T_a and T_s are relatively prime, the sequence just described must have the same linear complexity as the original a-sequence, so it has to satisfy a polynomial equation $Q(.) = 0$ of degree $|a|$. But then also the sequence $x_{[s]}$ has to satisfy this equation, i.e., $Q(x_{[s]}) = 0$ (with a little abuse of notation). Now define $P(z) := Q(z^{2^{|s|-1}})$. The polynomial P satisfies $P(z) = 0$ and has degree $|a| 2^{|s|-1}$, which is an upper bound for L.

2. Lower bound: Denote by $M(z)$ the minimal polynomial for the sequence x. Since $Q(x_{[s]}) = 0$, it follows that the polynomial $M(z)$ is a divisor of the polynomial

$$Q(x_{[s]}) = Q(z^{2^{2^{|s|-1}}}) = Q(z)^{2^{|s|-1}},$$

hence

$$M(z) = Q(z)^t$$

for some $t \leq 2^{|s|-1}$. Now assume that the lower bound asserted in Theorem 5.3 is not true and let $t \leq 2^{|s|-2}$. Then $M(z)$ divides $Q(z)^{2^{|s|-2}}$. Since $Q(z)$ is an irreducible polynomial of degree $|a|$, it divides the polynomial $1 + z^{T_a}$, so it follows that the polynomial $M(z)$ divides the polynomial

$$(1 + z^{T_a})^{2^{|s|-2}} = 1 + z^{T_a 2^{|s|-2}},$$

which entails that the period of the x-sequence can be at most $T_a 2^{|s|-2}$. But this is a contradiction to Theorem 5.2, hence we must indeed have $t > 2^{|s|-2}$. $\square$

The next theorem, which we state without proof, gives statistical properties of the shrinking generator. The general assertion is that one can show that the distribution of the output sequence of a shrinking generator is "near" to the distribution of a genuine unbiased random sequence in the following sense:

Theorem 5.4. *Consider a shrinking generator as above. Denote by U a genuine unbiased random sequence of length n. Let $b \in \{0, 1, *\}^n$ be any template. Then*

$$|E(template_b(X^{(n)})) - E(template_b(U)| = O(\frac{n}{2^{|a|}}).$$

Furthermore, assume $x_{(1)}$ and $x_{(2)}$ are two elements of the x-sequence with distance ℓ. Then the correlation between $x_{(1)}$ and $x_{(2)}$ is bounded by $O(\frac{\ell}{2^{\lceil a \rceil}})$.

(See Coppersmith et al. (1994), Theorem 13 and Corollary 14).

A concept related to the shrinking generator is the so-called self-shrinking generator. There, one works with only one LFSR and consecutive (non-overlapping) pairs of its output bits. If the first bit of the pair is a 1, then the second bit of the pair is included in the x-sequence (output) of the self-shrinking generator, otherwise the pair is dicarded. For more information about the self-shrinking generator see Meier, Staffelbach (1995) and Blackburn (1999). In the latter paper, the maximum linear complexity conjectured by Meier and Staffelbach (1995) is proven.

5.3 Perfect Pseudo-randomness

In this section, we give a definition of so-called "perfect" pseudo-randomness. Loosely speaking, this means a pseudo-random source that can not "efficiently" be distinguished by a computer from a truly random sequence. However, the test for perfect pseudo-randomness is not practically implementable. For a function $f(n)$ ($n \in I\!\!N$) we will write $f(n) = O(\nu(n))$ if $f(n) = O(1/g(n))$ ($n \to \infty$) for every polynomial $g(z)$. In this case, we will say that the function $f(n)$ is negligible. A model M is called a perfect simulation of a source S if for every probabilistic polynomial-time algorithm $D : I\!\!B^n \to I\!\!B$ we have

$$|P_S(D = 1) - P_M(D = 1)| = O(\nu(n)).$$

This means that no probabilistic polynomial algorithm can distinguish S from M with non-negligible probability, or, in other words, that S and M are polynomially indistinguishable. If D did not satisfy the above inequality, then we would say that D is a distinguishing algorithm. The following theorem states that the so-called Comparative Next Bit Test is a test of perfect pseudo-randomness. However, this test is of asymptotic nature and involves a formulation of type "for every polynomial-time algorithm", so it is only of theoretical value, since there are infinitely many such algorithms. But even in theory, up to now it is not yet known if perfect pseudo-random generators actually exist!

Definition 5.1. *A source S passes the Comparative Next Bit Test with respect to a model M if, for every $i \in \{1, 2, \ldots, n\}$ and every probabilistic polynomial-time algorithm $A : \mathbb{B}^{i-1} \to \mathbb{B}$, we have that*

$$|P_S(A(x^{(i-1)}) = x_i) - P_M(A(x^{(i-1)}) = x_i)| = O(\nu(n)).$$

Theorem 5.5. *A model M is a perfect simulation of a source S iff S passes the Comparative Next Bit Test with respect to M.*

Proof: The "only if"-direction is easy to see by contraposition. What is more difficult is the "if"-direction which we will verify in the following. Suppose S is not a perfect simulation of M. We have to prove that S does not pass the Comparative Next Bit Test with respect to M. Let $D : \mathbb{B}^n \to \mathbb{B}$ be a distinguishing algorithm, i.e.,

$$|P_S(D(x^{(n)}) = 1) - P_M(D(x^{(n)}) = 1)| \geq n^{-k}$$

for some constant exponent k. Let p_i^S denote the probability that the algorithm D gives 1 as output when the first i bits of its input are taken out of the source S and the rest are i.i.d. unbiased random bits. By replacing S by M in the above sentence, we define p_i^M analogously. Consider the difference $d_i := p_i^S - p_i^M$. It holds that $p_n^S = P_S(D(x^{(n)}) = 1)$, $p_n^M = P_M(D(x^{(n)}) = 1)$, $p_0^S = p_0^M = P_U(D(x^{(n)}) = 1)$ (where U means a source of genuine independent i.i.d. unbiased random bits). Thus as $d_0 = 0$ and $|d_n| = |p_n^S - p_n^M| \geq n^{-k}$, there must be an i such that $|d_i - d_{i-1}| \geq n^{-(k+1)}$. W.l.o.g. $d_i > 0$. The Comparative Next Bit Test A inputs in D the concatenated bitstring $x^{(n)} = x^{(i-1)} x_i^{(n)}$ (where $x^{(i-1)} \in S$ or M and $x_i^{(n)}$ is a bitstring generated by running the source U $n-i+1$ times). The output will be x_i if $D(x^{(n)}) = 1$ and $1 - x_i$ else. Now let $x_1, x_2, \ldots, x_i$ be bits produced by S or M and let q^S resp. q^M be the probability that the distinguisher D yields 1 as output when bits number $1, 2, \ldots, i-1$ are given by $x_1, x_2, \ldots, x_{i-1}$, bit number i is $1 - x_i$, and the rest are independent i.i.d. unbiased random bits. Then we have

$$p_{i-1}^S = \frac{p_i^S + q^S}{2},$$

$$p_{i-1}^M = \frac{p_i^M + q^M}{2},$$

and thus

$$P_S(A(x^{(i-1)}) = x_i) = \frac{1}{2}p_i^S + \frac{1}{2}(1 - q^S)$$
$$= \frac{1}{2} + p_i^S - p_{i-1}^S.$$

On the other hand

$$P_M(A(x^{(i-1)}) = x_i) = \frac{1}{2}p_i^M + \frac{1}{2}(1 - q^M)$$
$$= \frac{1}{2} + p_i^M - p_{i-1}^M,$$

hence

$$P_S(A(x^{(i-1)}) = x_i) - P_M(A(x^{(i-1)}) = x_i) \geq n^{-(k+1)}. \square$$

The property that the Comparative Next Bit Test checks can be called unpredictability, more precisely forwards unpredictability. Since the property of a pseudorandom number generator to be perfectly pseudorandom or not does not change if the output bits are taken in reverse order, forwards unpredictability is equivalent to backwards unpredictability.

A permutation $f(.)$ is called one-way, if its result can be calculated in polynomial time, but on the other hand, for any probabilistic polynomial-time algorithm A the probability $P(A(f(x)) = x)$ is negligible. A predicate (bit) $B(.)$ is called hard-core for the permutation $f(.)$ if $B(f(.))$ can be determined in polynomial time whereas for all probabilistic polynomial time algorithms A, the difference $P(A(x) = B(x)) - 1/2$ is neglibile. If there exists a hard-core bit, then the permutation has to be one-way. Blum and Micali (1984) have proved that every one-way permutation $f(.)$ with hard-core bit $B(.)$ gives rise to a perfect pseudorandom generator as follows:

Theorem 5.6. (Blum and Micali.) *Assume $f(.)$ is a one-way permuation with hard-core bit $B(.)$. Then the iteration*

$$g(x) = (B(f(x)), B(f(f(x))), B(f(f(f(x)))), \ldots)$$

yields a perfect pseudorandom generator.

Proof: We will show that the above generator is not backwards predictable. Assume the contrary. Then one could guess, in polynomial time and with non-neglibile probability of success, the value $B(f^n(x))$ (the n-fold iteration of f) given the set of values $S = \{B(f^m(x)) : m \geq n+1\}$. But S can be computed in polynomial time from $f^n(x)$. So one can guess in polynomial time and with probability of success non-negligibly greater than $1/2$ the value $B(f^n(x))$. So $B(.)$ can not be hard-core, since $f^n(x)$ has the same distribution as $f(x)$ by the fact that $f(.)$ is supposed to be a permutation. $\square$

In the Blum-Micali generator, x plays the role of a random seed. So these generators rather serve to improve randomness than produce it.

5.4 Local Statistics and de Bruijn Shift Registers

A general feedback shift register (FSR for short) of length n is a feedback shift register that is defined like an LFSR with the exception that the feedback function f needs not be linear, but can be an arbitrary function $f : B^n \to B$ (i.e., a polynomial in n Boolean variables of arbitrary degree).

Assume that $x = \{x_i\}_{i \geq 0}$ is an m-periodic bitsequence. We say that x has an [almost] ideal local statistics of order h if [almost] every h-tuple appears the same number of times as a subsequence of $x^{(m+h-1)}$. The following nested property of local statistics holds: If the m-periodic sequence x has [almost] ideal local statistics of order h, then it has also [almost] ideal local statistics of order $1, 2, \ldots, h - 1$.

Proposition 5.1. *An FSR of length n can not produce an output sequence with [almost] ideal local statistics of order $n + 1$.*

Proof: Any pattern of n consecutive bits in x determines uniquely the next bit. Thus at most 2^n of the possible 2^{n+1} patterns of n consecutive bits can occur in x, hence a fraction of at least $1 - 2^n/2^{n+1} = 1/2$ of the possible subsequences of length n never occur in x. So almost ideal local statistics of order $n + 1$ is not possible. $\square$

Now we want to characterize those FSR of length n that produce ideal local statistics of order n. Here the notion of a so-called de Bruijn FSR turns out to be crucial.

An FSR is called non-singular, if all its states lie on closed cycles in its state-transition diagram. Other states would be called transient states, so an FSR is non-singular iff it has no transient states. In other words, an FSR is non-singular iff every state has a unique predecessor state. A de Bruijn FSR is a non-singular FSR with only one cycle. Non-singularity can be be characterized algebraically by the Golomb-Welch Theorem:

Theorem 5.7. *(Golomb-Welch) An FSR of length n is non-singular iff its feedback function f satisfies*

$$f(x_{j-n}, \ldots, x_{j-1}) = x_{j-n} + g(x_{j-2}, \ldots, x_{j-n}). \tag{5.14}$$

Proof: We first prove that (5.14) is necessary for non-singularity. Relation (5.14) holds iff

$$f(0, a_2, a_3, \ldots, a_n) = 1 + f(1, a_2, a_3, \ldots, a_n)$$

for all $a_2, a_3, \ldots, a_n \in B$. If the FSR is singular, then there must exist at least one state $(b_1, \ldots, b_n) \in B^n$ with two predecessor states in the state-transition diagram. But then

$$f(0, b_1, b_2, \ldots, b_{n-1}) = f(1, b_1, b_2, \ldots, b_{n-1}),$$

hence (5.14) can not be fulfilled.

Now let us show sufficiency. Predecessors of state $(a_1, a_2, \ldots, a_n) \in I\!B^n$ have the form $(b, a_1, a_2, \ldots, a_{n-1}) \in I\!B^n$ with

$$a_n = f(b, a_1, a_2, \ldots, a_{n-1}) = b + g(a_1, a_2, \ldots, a_{n-1}). \tag{5.15}$$

Equation (5.15) has the unique solution

$$b = a_n - g(a_1, a_2, \ldots, a_{n-1}),$$

hence every state has a unique predecessor. □

Corollary 5.2. *The period of the output sequence of an FSR of length n is at most 2^n with equality iff the FSR is a de Bruijn FSR. (In this case, the output sequence wiil be called a de Bruijn sequence.)*

Now the following property holds:

Theorem 5.8. *The output sequence s of an FSR of length n is a periodic sequence with ideal local statistics of order n iff the FSR is a de Bruijn FSR.*

Proof: Since there are 2^n different n-tuples and since the output sequence of an FSR has period $m \leq 2^n$, every n-tuple can appear in the subsequence $x^{(m+n-1)}$ only if $m = 2^n$. On the other hand, if $m = 2^n$, then every n-tuple has to appear exactly once in $x^{(m+n-1)}$. Hence $m = 2^n$ is a necessary and sufficient condition for x to have ideal local statistics of order n. But $m = 2^n$ means that the FSR is a de Bruijn FSR. □

5.5 Correlation Immunity

Here, the model of the key generator is the following: There are m LFSR yielding the outputs $\{x_i^{(j)}\}_{i \geq 0}$ ($1 \leq j \leq m$). These LFSR streams are gathered by a non-linear combining function $f : I\!B^m \to I\!B$ to yield the output

$$z_i = f(x_i^{(1)}, x_i^{(2)}, \ldots, x_i^{(m)}).$$

On a "short time" basis, the LFSR outputs may be well modeled by n independent symmetric binary sources. Now take for example $m = 3$ and

$$f(x_1, x_2, x_3) := x_1 x_2 + x_1 x_3 + x_2 x_3.$$

One sees that if the LFSR are modeled as above, then $P(z_i = 0) = P(z_i = 1) = 1/2$. But, if the LFSR 1 (i.e., that which generates the sequence $\{x_i^{(1)}\}_{i \geq 0}$) is known, we can mount the following correlation attack to find the true phase of LFSR 1 and thus find its initial state: If we multiply $\{z_i\}_{i \geq 0}$ with the shifted sequence $\{x_i^{(1)}\}_{i \geq 0}$ in the "correct" phase, then we see from the definition of f that the 1 occurs with probability $3/8$ instead of just $1/4$ as it would be with true binary symmetric random sequences. This type of attack can be done for every LFSR.

So a natural question is how to choose the combining function f to avoid such attacks.

Definition 5.2. *A function $f : \mathbb{B}^m \to \mathbb{B}$ is called h-th order correlation immune if, whenever $X_1, X_2, \ldots, X_m$ are independent unbiased $\mathbb{B}$-valued random variables, then $Z := f(X_1, X_2, \ldots, X_m)$ is independent of all finite subsequences $(X_{i_1}, X_{1_2}, \ldots, X_{i_h})$ ($1 \le i_1 \le \ldots \le i_h \le m$).*

The signification of h-th order correlation immunity lies in the fact that if a non-linear combination function f is h-th order correlation immune, then it is not possible to mount a correlation attack on any combination of h input sequences.

For a function $f : \mathbb{B}^m \to \mathbb{B}$, its Fourier (or Walsh-Hadamard) transform is defined as

$$F(\omega) := \sum_{x \in \mathbb{B}^m} f(x)(-1)^{\langle x, \omega \rangle} \qquad (\omega \in \mathbb{B}^m).$$

One has the following inversion formula:

$$f(x) := 2^{-m} \sum_{\omega \in \mathbb{B}^m} F(\omega)(-1)^{\langle x, \omega \rangle}.$$

Now correlation immunity can be characterized in terms of Fourier transforms as follows:

Theorem 5.9. *(Xiao-Massey Spectral Test) The function $f : \mathbb{B}^m \to \mathbb{B}$ is h-th order correlation immune iff its Fourier transform F satisfies*

$$F(\omega_1, \omega_2, \ldots, \omega_m) = 0$$

for all $\omega = (\omega_1, \omega_2, \ldots, \omega_m) \in \mathbb{B}^m$ with $1 \le w_H(\omega) \le h$ (where $w_H(\omega)$ denotes the Hamming weight (i.e. the number of entries 1) of the vector ω).

The proof of Theorem 5.9 follows from the following two lemmas:

Lemma 5.7. *Let X be a random vector consisting of m independent unbiased $\mathbb{B}$-valued random variables $X_1, X_2, \ldots, X_m$, $f : \mathbb{B}^m \to \mathbb{B}$, $\omega \in \mathbb{B}^m \backslash \{0\}$ and put $Z := f(X_1, X_2, \ldots, X_m)$. Then Z is independent of $\langle X, \omega \rangle$ iff $F(\omega) = 0$.*

Proof: Since

$$P_{Z|\langle X, \omega \rangle}(1|b)$$
$$= \frac{|\{x \in \mathbb{B}^m : f(x) = 1, \langle x, \omega \rangle = b\}|}{|\{x \in \mathbb{B}^m : \langle x, \omega \rangle = b\}|}$$
$$= 2^{-(m-1)} \sum_{x \in \mathbb{B}^m : \langle x, \omega \rangle = b} f(x),$$

we get

$$P_{Z|\langle X, \omega \rangle}(1|0) - P_{Z|\langle X, \omega \rangle}(1|1)$$
$$= 2^{-(m-1)} \sum_{x \in \mathbb{B}^m} f(x)(-1)^{\langle x, \omega \rangle}$$
$$= 2^{-(m-1)} F(\omega). \square$$

Lemma 5.8. *A discrete random variable Z is independent of the random vector $Y = (Y_1, Y_2, \ldots, Y_m) \in \mathbb{B}^m$ iff for every $a \in \mathbb{B}^m$, Z is independent of $\langle Y, a \rangle$.*

The proof follows directly from considering Fourier transforms (see Brynielsson (1989)). $\square$

We point out that Theorem 5.9 is really applicable in practice, since to compute the Fourier transform needs at most $O(m2^m)$ additions and subtractions (see Massey (1997), p. 3.63). However, high-order correlation immunity cannot happen if the nonlinear order λ of the function f is too high. Let us explain this in detail: Let $f : \mathbb{B}^m \to \mathbb{B}$. Then the so-called algebraic normal form of the function f is

$$
\begin{aligned}
f(x_1, x_2, \ldots, x_m) = {} & a_0 + a_1 x_1 + a_2 x_2 + \ldots + a_m x_m \\
& + a_{1,2} x_1 x_2 + a_{1,3} x_1 x_3 + \ldots \\
& + \ldots \\
& + a_{1,2,\ldots,m} x_1 x_2 \ldots x_m,
\end{aligned}
$$

where the coefficients are given by the inversion formula

$$
a_{1,2,\ldots,k} = \sum_{x \in S_{1,2,\ldots,k}} f(x_1, x_2, \ldots, x_m) \tag{5.16}
$$

and

$$
S_{1,2,\ldots,k} := \begin{cases} \{x : x_{k+1} = x_{k+2} = \ldots = x_m = 0\} & : \quad 1 \le k \le m-1 \\ \{x\} & : \quad k = m, \end{cases} \tag{5.17}
$$

etc. (see Siegenthaler (1984)).

Definition 5.3. *The nonlinear order λ of a function $f : \mathbb{B}^m \to \mathbb{B}$ is the maximum number of variables x_j that occur in a term of the algebraic normal form of f.*

Theorem 5.10. *(Siegenthaler's Inequality) If λ denotes the nonlinear order of the function $f : \mathbb{B}^m \to \mathbb{B}$ and if f is h-th order correlation immune, then*

$$
h \le m - \lambda.
$$

Proof: Assume f is h-th order correlation immune for some $h \in \{1, 2, \ldots, m-1\}$. We show that no product of $m - h + 1$ or more variables x_j can occur in the algebraic normal form of f. Define the numbers

$$
N_{1,2,\ldots,k} = |\{x \in \mathbb{B}^m : x \in S_{1,2,\ldots,k}, f(x) = 1\}|. \tag{5.18}
$$

Let $Z := f(X)$ where X is a vector of m independent unbiased $\mathbb{B}$-valued random variables. Then we get

$$P(Z = 1 \mid X_{k+1} = X_{k+2} = \ldots = X_m = 0)$$
$$= \frac{N_{1,2,\ldots,k}}{2^k} \quad (1 \le k \le m - 1) \tag{5.19}$$

and

$$P(Z = 1) = \frac{N_{1,2,\ldots,m}}{2^m}. \tag{5.20}$$

We obtain

$$P(Z = 1 \mid X_{k+1} = X_{k+2} = \ldots = X_m = 0)$$
$$= P(Z = 1) \quad (m - h \le k \le m - 1)$$

and hence from (5.19) and (5.20)

$$\frac{N_{1,2,\ldots,m}}{2^m} = \frac{N_{1,2,\ldots,m-1}}{2^{m-1}} = \ldots = \frac{N_{1,2,\ldots,m-h}}{2^{m-h}},$$

which implies

$$N_{1,2,\ldots,k} = 2^{k-(m-h)} N_{1,2,\ldots,m-h} \quad (m - h \le k \le m). \tag{5.21}$$

From (5.21), for $m - h + 1 \le k \le m$, these numbers must be even, which implies, from (5.16) and (5.17)

$$a_{1,2,\ldots,k} = 0 \quad (m - h + 1 \le k \le m).$$

However, this argument not only applies to the first k components of x, but to any k components of x, which proves the assertion. $\square$

The tradeoff given by Siegenthaler's Inequality does not exist if the combining function f is allowed to have memory. We will not persue this track further and only refer to Rueppel (1986), Chapter 9.

Further seminal papers on correlation attacks are e.g. Chepyzhov, Smeets (1991) and Meier, Staffelbach (1989), (1991), (1992).

5.6 The Quadratic Congruential Generator

Now we will consider a special example of the above Blum-Micali generator, namely the quadratic congruential generator. Its implementation can be done by a (especially simple) non-linear shift register of length 1. Assume n is a Blum integer, i.e., a product of two distinct odd primes p and q both congruent to 3 (mod.4). (In particular, the factoring of Blum integers is believed to be computationally hard.) Let k be the length of the binary expansion of n ($k := |n|$). For an integer x, define the "absolute value" mod.n by

$$|x|_n := \begin{cases} x(mod.n) & : \quad x(mod.n) < n/2 \\ n - (x(mod.n)) & : \quad x(mod.n) \ge n/2. \end{cases}$$

Then take, as permutation $f(.) = f_n(.)$, the "absolute value" of the square:

$$f_n(x) := |x^2 (mod.n)|_n.$$

(By Euler's criterion (mentioned in Section 2.2) for prime factors of Blum integers we have that

$$\begin{aligned}((n - y)|n) &= (-y|n) \\ &= (-1|n)(y|n) \\ &= (-1|p)(-1|q)(y|n) \\ &= (y|n).\end{aligned}$$

Hence exactly one square root of a quadratic residue modulo n is less than $n/2$, since every quadratic residue modulo n has exactly two square roots with Legendre-Jacobi symbol equal 1. So f is really a permutation of the set

$$S = \{x \in \mathbb{Z}_n^* : 0 \le x < n/2, (x|n) = 1\}.)$$

Using this permutation in the Blum-Micali construction will be called the quadratic congruential generator. The main aim of this section will be to prove and discuss the following important fact:

Theorem 5.11. *Breaking the quadratic congruential generator is probabilistic polynomial-time equivalent to the factoring of n.*

First we will show the following:

Theorem 5.12. *Inverting $f_n(.)$ is probabilistic polynomial-time equivalent to factoring $n = pq$ (p, q primes).*

Proof: On given as input a square $z = y^2 (mod.n)$ with $0 < z < n/2$ and a square root y with $(y/n) = -1$ is known, then a probabilistic polynomial-time algorithm A that inverts $f_n(.)$ will output a square root x of z in the domain of $f_n(.)$ with probability greater than, say, $1/\delta(|n|)$ (where $|n|$ is the length of the binary expansion of n and δ is some polynomial). Since $(x/n) = 1$ (by the definition of the domain of f_n), but also $(-1/n) = 1$, it is not possible that $x = \pm y (mod.n)$. But

$$pq|x^2 - y^2 = (x - y)(x + y),$$

which entails that exactly one of the two primes p, q divides $x + y$ evenly. From the Euclidean algorithm one can compute $\gcd(n, x + y)$ in polynomial time, which yields the factors p and q. The probability that a randomly chosen $y \in \mathbb{Z}_n^*$ satisfies indeed $(y/n) = -1$ and also the probability that $0 < z < n/2$ are both $1/2$. If y is a randomly (with uniform distribution) selected element of the set $\{y \in \mathbb{Z}_n^* : (y/n) = -1\}$, then also z is uniformly distributed on the set of all quadratic residues mod.n. So one can run the following probabilistic algorithm: Generate at random (with uniform distribution) an element

$y \in \mathbb{Z}_n^*$. Repeat this until for $z := y^2 (mod.n)$ one has $0 < z < n/2$. Then input z to the inverting algorithm A. Check if $A(z)^2 = z(mod.n)$. The mean number of times this has to be done until one finds a square root that allows to factor n is $2\delta(|n|)$. If A is polynomial, the whole calculation proceeds is expected polynomial-time. □

Proof of Theorem 5.11: 1. The strategy of proof will be to show that $\mathrm{Lsb}(f_n^{-1}(.))$, the least significant bit of $f_n^{-1}(.)$ is hard-core and then to use Theorem 5.12. In other words, we will prove that under the hypothesis that a probabilistic polynomial-time algorithm that can guess the least significant bit of $f_n^{-1}(.)$ with probability non-negligibly greater that $1/2$, then one can construct a probabilistic polynomial-time algorithm for inverting $f_n(.)$. So there is some similarity to Section 2.4. Let $\mathcal{O}$ denote an oracle that takes as input n and an x in the range of $f_n(.)$ and yields as output a guess for $\mathrm{Lsb}(f_n^{-1}(.))$ that is correct with probability $1/2 + |n|^{-c}$ for some constant c. Now the method for constructing a probabilistic polynomial-time algorithm for inverting $f_n(.)$ will be to call the oracle $\mathcal{O}$ at most polynomially many times to find $f_n^{-1}(.)$ with the aid of a gcd-algorithm that makes all its computations based solely on the least significant bits of all involved integers (so that we can use $\mathcal{O}$). This can be done with the Brent-Kung algorithm, which we will describe later and from which we will show that indeed with a probability lower-bounded by the inverse of some polynomial in k yields the correct answer, so that the experiment described in the following needs to be repeated only an expected number of times that is polynomial in l. Using the Brent-Kung algorithm for calculating greatest common divisors, we compute, for randomly chosen a, b, the greatest common divisor $\gcd([ax]_n, [bx]_n)$ based on the permuted values $f_n(ax(mod.n))$ and $f_n(bx(mod.n))$, where

$$[z]_n := \begin{cases} z(mod.n) & : & z(mod.n) < n/2 \\ z(mod.n) - n & : & z(mod.n) \geq n/2. \end{cases}$$

When we have finished the Brent-Kung algorithm, we will be in possession of a representation of $[dx]_n := \gcd([ax]_n, [bx]_n)$ of the latter gcd, hence d and $f_n(dx(mod.n)) = f_n([dx]_n)$ are known. If $[ax]_n$ and $[bx]_n$ are relatively prime (an event whose probability tends asymptotically to $6/\pi^2$ as $n \to \infty$ due to a theorem of Dirichlet), then it follows that

$$[dx]_n = \pm 1 \tag{5.22}$$

and therefore $f_n(dx) = 1$. If we check $f_n(x) \stackrel{?}{=} f_n(\pm d^{-1}(mod.n))$ (note that the Euclidean algorithm calculates inverses in polynomial time without knowing the prime factorization of n) and find that these two values are indeed equal, then we have a good probability that one of the values $\pm d^{-1}(mod.n)$ lies indeed in the domain of $f_n(.)$ and we are finished. Otherwise, repeat the experiment sufficiently (probabilistically polynomially) many times.

2. Now we turn to the description of the Brent-Kung algorithm:
Given two integers A (odd) and B with lengths $\leq |n|$, repeat the following steps until $B = 0$:

- While $\text{Lsb}(|B|) = 0$, do $B \leftarrow B/2$; $\text{length}(B) \leftarrow \text{length}(B) - 1$.
- If $\text{length}(B) < \text{length}(A)$, then $\text{swap}(A, B)$; $\text{swap}(\text{length}(A), \text{length}(B))$.
- If $\text{Lsb}(|(A + B)/2|) = 0$, then $B \leftarrow (A + B)/2$; else $B \leftarrow (A - B)/2$.

(If A is even, then in step 3 the expression $\text{Lsb}((A + B)/2)$ makes no sense. But in this case we can evidently reduce A before, so that it will become odd.)
(This Brent-Kung algorithm rests on the following facts:
(i) If a, b are both even, then $\gcd(a, b) = \gcd(a/2, b/2)$.
(ii) If a is odd and b is even, then $\gcd(a, b) = \gcd(a, b/2)$.
(iii) If a, b are both odd, then $\gcd(a, b) = \gcd(a, (a+b)/2) = \gcd(a, (a-b)/2)$.)
After halting of the Brent-Kung algorithm (i.e. $B = 0$), the variable (memory cell) A contains the gcd of the two original input numbers A and B. One counts that the maximal number of evaluations of a least significant bit in the Brent-Kung algorithm is $O(|n|)$.
3. Let us now explain how the Brent-Kung algorithm is used in our problem. In our application, we must put $A := [ax]_n$ and $B := [bx]_n$ and the algorithm will work only with $f_n(ax(mod.n))$ and $f_n(bx(mod.n))$ with the aid of the oracle $\mathcal{O}$. We first define the so-called parity by

$$\text{par}(bx(mod.n)) := \text{Lsb}(|B|) = \text{Lsb}(|[bx]_n|). \tag{5.23}$$

The bit $\text{Lsb}(|B|)$ is what we really want to know at the end, so from (5.23) we must look for a procedure that calculates the parity, using the oracle $\mathcal{O}$.
4. The parity algorithm (sketch): The basic principle here is to determine $\text{par}(dx(mod.n))$ by comparing the $\text{Lsb}(s)$ with $\text{Lsb}(s + dx(mod.n))$ for randomly chosen $s \in \mathbb{Z}^*_{\lfloor n/2 \rfloor}$ (with uniform distribution). If no "wraparound 0" occurs by adding s to dx, then one has the relation

$$\text{par}_n(dx(mod.n)) = \text{Lsb}(s) + \text{Lsb}(s + dx(mod.n)) \quad (mod.2). \tag{5.24}$$

The probability of a "wraparound 0" can be shown to be small. If one chooses s at random as described above, then unfortunately the values of $(s + dx)(mod.n)$ are not known. How to overcome this difficulty and for further details, in particular the control of possible errors, we refer to Brands, Gill (1996). $\square$

6 An Information Theory Primer

6.1 Entropy and Coding

In this section, we will introduce one of the most important notions in cryptology, namely the information content and the entropy. The entropy of a random variable X that can assume the n different values $x_1, x_2, \ldots, x_n$ with the respective probabilities $p_1, p_2, \ldots, p_n$ is defined as

$$H(X) = H(p_1, p_2, \ldots, p_n) = \sum_{i=1}^{n} p_i \log_2(1/p_i) = -\sum_{i=1}^{n} p_i \log_2 p_i.$$

If one considers, e.g., a decision tree, then one sees easily that $\log_2(1/p_i)$ (unit: "bit") may be interpreted as the information content of the realization x_i of the random variable X, i.e. the entropy is the average information content of a realization of X. Actually, the entropy only depends on the probabilites $p_1, p_2, \ldots, p_n$ and not of the realizations $x_1, x_2, \ldots, x_n$ themselves. If we define

$$\mathcal{X} := \{x_1, x_2, \ldots, x_n\}$$

as the alphabet, then, instead of $H(X)$, we will sometimes also write $H(\mathcal{X})$. In coding theory, the entropy is (loosely speaking) the optimal average length of a codeword in an alphabet with two letters, as we will see in the following.

Lemma 6.1. *If for $p_i, q_i > 0$ we have the inequality $\sum_{i=1}^{n} q_i \leq \sum_{i=1}^{n} p_i$, then it follows that*

$$-\sum_{i=1}^{n} p_i \log_2 p_i \leq -\sum_{i=1}^{n} p_i \log_2 q_i.$$

Proof: Since $\log x \leq x - 1$ we have $\log_2 x \leq \frac{x-1}{\log 2}$, hence

$$\sum_{i=1}^{n} p_i \log_2(q_i/p_i) \leq \frac{1}{\log 2} \sum_{i=1}^{n} p_i((q_i/p_i) - 1)$$

$$\leq \frac{1}{\log 2} \left(\sum_{i=1}^{n} q_i - \sum_{i=1}^{n} p_i\right)$$

$$\leq 0. \square$$

D. Neuenschwander: Prob. and Stat. Methods in Cryptology, LNCS 3028, pp. 77–88, 2004.
© Springer-Verlag Berlin Heidelberg 2004

If in the above lemma we put $q_i = 1/n$ $(i = 1, 2, \ldots, n)$, then the inequality

$$H(p_1, p_2, \ldots, p_n) \leq H(\frac{1}{n}, \frac{1}{n}, \ldots, \frac{1}{n}) = \log_2 n$$

follows. Hence the entropy of a random variable with n possible values becomes maximal if these n values are uniformly distributed, and in this case it assumes the value $\log_2 n$.

Given an alphabet $\mathcal{X} = \{x_1, x_2, \ldots, x_n\}$ with the n different letters $x_1, x_2, \ldots, x_n$ (so, e.g., for the latin alphabet we have $n = 26$ und $x_1 = "A", x_2 = "B", \ldots, x_{26} = "Z"$), then a (binary) encoding is a map

$$c : x_i \mapsto c_i,$$

which assigns a (finite) bitsequence c_i to every letter x_i from the alphabet $\mathcal{X}$ such that the condition of decodability (or "unique decoding") is fulfilled: Two different sequences of letters of plaintext must yield different codes. A sharper condition is irreducibility (Fano condition): No codeword is allowed to be the beginning of another codeword. Let ℓ_i denote the length of the codeword for the letter x_i. If p_i is the probability of occurrence of letter x_i, then

$$\bar{\ell} = \sum_{i=1}^{n} p_i \ell_i$$

is the average length of a codeword.

Lemma 6.2. (Kraft's inequality) *For a decodable code we have*

$$\sum_{i=1}^{n} 2^{-\ell_i} \leq 1.$$

Proof: The proof uses generating functions of sequences. Let a_k be the number of letters from the alphabet $\mathcal{X}$ with a codeword of length k and b_m the number of sequences of letters (from $\mathcal{X}$) with a codeword of length m. Then we have

$$b_m = \sum_{k=1}^{m} a_k b_{m-k}$$

(where $b_0 := 1$). If we consider the generating functions of $\{b_m\}_{m \geq 1}$ and $\{a_k\}_{k \geq 1}$

$$g(z) = \sum_{m=1}^{\infty} b_m z^m \quad (|z| < 1/2)$$

(convergent since $b_m \leq 2^m$ due to the "unique decoding condition") and

$$f(z) = \sum_{k=1}^{\infty} a_k z^k \quad (z \in \mathbb{R})$$

(polynomial!), then we obtain

$$g(z) = \sum_{m=1}^{\infty} b_m z^m$$
$$= \sum_{k=1}^{\infty} \sum_{m=1}^{\infty} a_k b_{m-k} z^m$$
$$= \sum_{k=1}^{\infty} a_k z^k \sum_{i=0}^{\infty} b_i z^i$$
$$= f(z)(g(z) + 1)$$

(where we have put $i = m - k$). Hence

$$f(z) = \frac{g(z)}{g(z) + 1}.$$

Since $g(z) \geq 0$ ($0 \leq z < 1/2$), we have that the polynomial $f(z) \leq 1$ ($0 \leq z < 1/2$), hence $f(1/2) \leq 1$ due to continuity, which yields the assertion. $\square$
On the other hand, it holds that

Lemma 6.3. *For given $\ell_1, \ell_2, \ldots, \ell_n \in \mathbb{N}$ that obey Kraft's inequality, there exists an irreducible code such that the codeword c_i is of length ℓ_i.*

Proof: The proof is of combinatorial nature. Let I_k be the set of all $i \in \{1, 2, \ldots, n\}$ such that $\ell_i = k$ and denote by a_k the number of elements of I_k. Thus the letters x_i for which $i \in I_1$ can be encoded by codewords of length 1. Now we proceed by recursion. Assume all letters having a codeword of maximal length $k - 1$ are encoded. Then one has used, for $j < k$, always a_j bitsequences of length j that are initial strings of always 2^{k-j} bitsequences of length k. Thus under the condition of irreducibility,

$$2^k - \sum_{j=1}^{k-1} a_j 2^{k-j} = 2^k \left(1 - \sum_{j=1}^{k-1} a_j 2^{-j}\right)$$

bitsequences still remain at our disposal. Since Kraft's inequality must hold, the latter quantity is $\geq 2^k(a_k 2^{-k}) = a_k$, so that enough codewords remain for also encoding all letters x_i with $i \in I_k$. $\square$
Let us assume that the letters x_i of the alphabet $\mathcal{X}$ occur with the respective probabilities p_i and put $H(\mathcal{X}) = H(p_1, p_2, \ldots, p_n)$.

Theorem 6.1. (Coding Theorem) *(i) Every decodable encoding of the alphabet $\mathcal{X}$ has an average codeworth length*

$$\bar{\ell} \geq H(\mathcal{X}).$$

(ii) On the other hand there exists an irreducible code with average codeword length

$$\bar{\ell} \leq H(\mathcal{X}) + 1.$$

Proof: (i) From Kraft's inequality and Lemma 6.1 (by putting $q_i = 2^{-\ell_i}$) it follows that

$$H(\mathcal{X}) = -\sum_{i=1}^{n} p_i \log_2 p_i$$

$$\leq -\sum_{i=1}^{n} p_i \log_2(2^{-\ell_i})$$

$$= \sum_{i=1}^{n} p_i \ell_i$$

$$= \bar{\ell}.$$

(ii) Let $\ell_i = -\lfloor \log_2 p_i \rfloor$. Then we have

$$2^{-\ell_i} \leq 2^{-\log_2 p_i} = p_i,$$

which entails Kraft's inequality. Now Lemma 6.3 yields the existence of an irreducible code with these ℓ_i. On the other hand, one sees that

$$\bar{\ell} = \sum_{i=1}^{n} p_i \ell_i$$

$$\leq \sum_{i=1}^{n} p_i(-\log_2 p_i + 1)$$

$$= H(\mathcal{X}) + 1.\square$$

To find such an optimal code explicitly is another problem of coding theory, which will not be considered here. It has to do with so-called Huffman trees, which are rooted binary trees all of whose non-leaves are arranged, from left to right, in order of non-decreasing distance from the root.

Another (less known) complexity measure is the so-called marginal guesswork. It has nothing to do with the entropy in the sense that there are no general inequalities relating these two measures onto another. This approach will be presented in Section 6.3.

6.2 Relative Entropy, Mutual Information, and Impersonation Attack

In this section, we will collect some results about information theory involving several probability measures. First, we define the "relative entropy" between two probability measures:

Definition 6.1. *Let P and Q be two probability measures on the same alphabet $\mathcal{X}$. Then the relative entropy (information divergence, Kullback-Leibler distance, discrimination) from P to Q is defined as*

$$D(P||Q) := -\sum_{x \in S} P(x) \log_2 \frac{Q(x)}{P(x)}.$$

Note that in general $D(P||Q) \neq P(Q||P)$. The following lemma shows that the word "distance" is chosen reasonably.

Lemma 6.4. $D(P||Q) \geq 0$ with equality iff $P = Q$.

Proof: From the inequality $\log r \leq r - 1$ with equality iff $r = 1$ we deduce

$$D(P||Q) \geq -\frac{1}{\log 2} \sum_{x \in \mathcal{X}} P(x)(\frac{Q(x)}{P(x)} - 1)$$

$$= \frac{1}{\log 2}(-\sum_{x \in \mathcal{X}} Q(x) + \sum_{x \in \mathcal{X}} P(x))$$

$$\geq 0,$$

with equality iff $P(x) = Q(x)$ for all $x \in \mathcal{X}$. $\square$

The relative entropy has important properties for hypothesis testing. This will be used, e.g., in connection with the impersonation attack presented at the end of this section. Consider the null hypothesis H_0 and the alternative H_1 and let V be the decision: $V = 0$ if we decide for H_0 and $V = 1$ if we decide for H_1. Let $\mathcal{D}_0$ be the decision region for H_0 and $\mathcal{D}_1$ that for H_1. Consider the probability measures

$$P := P_{Y|H_0},$$

$$Q := P_{Y|H_1}.$$

One can interpret the relative entropy from P to Q as the expectation of the log-likelihood ratio under the null hypothesis:

$$D(P||Q) = E(\log_2 \frac{P(Y)}{Q(Y)} \mid H_0)$$

(since

$$D(P||Q) = D(P_{Y|H_0}||P_{Y|H_1})$$

$$= -\sum_{y \in \mathrm{supp}(P_{Y|H_0})} P_{Y|H_0}(y) \log_2 \frac{P_{Y|H_1}(y)}{P_{Y|H_0}(y)}).$$

Let α, resp. β, denote the drror probabilites of first, resp. second, kind:

$$\alpha := P(Y \in \mathcal{D}_1|H_0) = \sum_{y \in \mathcal{D}_1} P(y),$$

$$\beta := P(Y \in \mathcal{D}_0|H_1) = \sum_{y \in \mathcal{D}_0} Q(y).$$

Then we also have

$$D(P_{V|H_0}||P_{V|H_1}) = -\alpha \log_2 \frac{1-\beta}{\alpha} - (1-\alpha) \log_2 \frac{\beta}{1-\alpha}.$$

Theorem 6.2. *It holds that*

$$D(P_{V|H_0}||P_{V|H_1}) \le D(P||Q),$$

with equality iff for the likelihood ratio $L(y) := P(y)/Q(y)$ we have $L(y) = \ell_0$ for all $y \in \mathcal{D}_0$ and $L(y) = \ell_1$ for all $y \in \mathcal{D}_1$.

Proof: Consider the events $A_i = \{V = i\}$ ($i = 0, 1$). We observe

$$D(P||Q) = E(-\log_2 \frac{Q(Y)}{P(Y)}|H_0)$$

$$= E(-\log_2 \frac{Q(Y)}{P(Y)}|H_0 \cap A_0)P(A_0|H_0)$$

$$+ E(-\log_2 \frac{Q(Y)}{P(Y)}|H_0 \cap A_1)P(A_1|H_0). \qquad (6.1)$$

However, $P_{Y|H_0 \cap A_0}(y) = P(y)/(1-\alpha)$ if $y \in \mathcal{D}_0$ and 0 if $y \in \mathcal{D}_1$. So, by the concavity of the log-function, it follows that

$$E(-\log_2 \frac{Q(Y)}{P(Y)}|H_0 \cap A_0) \ge -\log_2(E(\frac{Q(Y)}{P(Y)}|H_0 \cap A_0))$$

$$= -\log_2 \sum_{y \in \mathcal{D}_0} \frac{Q(y)}{P(y)} \frac{P(y)}{1-\alpha}$$

$$= -\log_2 \frac{\beta}{1-\alpha}.$$

Also $P(A_0|H_0) = 1 - \alpha$. By applying similar considerations for the second summand in (6.1) we obtain the asserted inequality. $\square$

An important special case of the above theorem is the following estimate: If we choose the error probability of the first kind to be 0 (this is the usual assumption in cryptography; we do not want that an honest cryptogram of Alice will be thought of as fraudulent by Bob), then it follows that the error probability of the second kind has the following lower bound:

Corollary 6.1. *If $\alpha = 0$, then*

$$\beta \ge 2^{-D(P||Q)}.$$

Now we turn to the definition of mutual information. For this, we first define the conditional entropy of the random variable X given Y:

$$H(X|Y) := E(-\log_2 P_{X|Y}(X|Y))$$

$$= -\sum_{(x,y) \in \mathrm{supp}(P_{(X,Y)})} P_{(X,Y)}(x,y) \log_2 P_{X|Y}(x|y).$$

Let us collect some properties of the conditional entropy in the form of lemmas. They can be proved based on the relative entropy.

Lemma 6.5. *It holds that*

$$0 \le H(X|Y) \le H(X)$$

with equality on the left-hand side iff Y uniquely determines X and with equality on the right-hand side iff X and Y are independent.

Proof: Let $P(x, y) := P_{(X,Y)}(x, y)$ and $Q(x, y) = P_X(x)P_Y(y)$. Then we have

$$0 \le D(P\|Q)$$
$$= - \sum_{(x,y) \in \mathrm{supp}(P_{(X,Y)})} P_{(X,Y)}(x, y) \log_2 \frac{P_X(x)P_Y(y)}{P_{(X,Y)}(x, y)}$$
$$= - \sum_{(x,y) \in \mathrm{supp}(P_{(X,Y)})} P_{(X,Y)}(x, y) \log_2 \frac{P_X(X)P_Y(y)}{P_{(X|Y)}(x|y)P_Y(y)}$$
$$= E(- \log_2 P_X(x) + \log_2 P_{X|Y}(X|Y))$$
$$= H(X) - H(X|Y)$$

with equality iff $P_{(X,Y)}(x, y) = P_X(x)P_Y(y)$ for all $(x, y) \in \mathrm{supp}(P_{(X,Y)})$, i.e. iff X and Y are independent. $\square$

Lemma 6.6.

$$H((X, Y)) = H(X) + H(Y|X).$$

Proof:

$$H((X, Y)) = E(- \log_2 P_{(X,Y)}(X, Y))$$
$$= E(- \log_2(P_X(X)P_{Y|X}(Y|X)))$$
$$= E(- \log_2 P_X(X)) + E(- \log_2 P_{Y|X}(Y|X))$$
$$= H(X) + H(Y|X). \square$$

Lemma 6.7.

$$H(X|(Y, Z)) \le H(X|Y)$$

with equality iff X and Z are independent, given that Y is known.

Proof: Put

$$P(x, y, z) := P_{(X,Y,Z)}(x, y, z)$$

and

$$Q(x, y, z) := P_Y(y)P_{X|Y}(x|y)P_{Z|Y}(z|y).$$

Then for the relative entropy we have

$$D(P\|Q) = -\sum_{(x,y,z)\in\mathrm{supp}(P_{(X,Y,Z)})} P_{(X,Y,Z)}(x,y,z)\log_2 \frac{P_Y(y)P_{X|Y}(x|y)P_{Z|Y}(z|y)}{P_{(X,Y,Z)}(x,y,z)}$$

$$= E(-\log_2 \frac{P_Y(Y)P_{X|Y}(X|Y)P_{Z|Y}(Z|Y)}{P_{(X,Y,Z)}(X,Y,Z)}).$$

But, since

$$P_{(X,Y,Z)} = P_Y(Y)P_{X|Y}(X|Y)P_{Z|(X,Y)}(Z|(X,Y)),$$

it follows that

$$D(P\|Q) = E(-\log_2 \frac{P_{Z|Y}(Z|Y)}{P_{Z|(X,Y)}(Z|(X,Y))})$$
$$= H(Z|Y) - H(Z|(X,Y))$$
$$\geq 0,$$

with equality iff for all x, y, z

$$P_Y(y)P_{(X,Z)|Y}((x,z)|y) = P_Y(y)P_{X|Y}(x|y)P_{Z|Y}(z|y),$$

which yields the assertion. $\square$

This allows us to define the mutual information $I(X;Y) := H(X) - H(X|Y)$; this is the information that Y gives about X. So Lemma 6.5 can be rewritten in the form

$$0 \leq I(X;Y) \leq H(X),$$

with equality on the left-hand side iff X and Y are independent and equality on the right-hand side iff Y uniquely determines X. More generally, one can define

$$I(X;Y|Z) := H(X|Z) - H(X|(Y,Z))$$

as the information that Y gives about X if Z is known. Then one also has

$$0 \leq I(X;Y|Z) \leq H(X|Z),$$

with equality on the left-hand side iff X and Y are independent given Z and equality on the right-hand side iff Y uniquely determines X if Z is given.

In cryptology, there is not only the problem of keeping a message secret, but that of Bob being able to be "reasonably" sure that he really gets what Alice has sent to him without Eve having changed the text. We underline that secrecy and authenticity/integrity are different properties, neither implies the other automatically.

One speaks of an impersonation attack in the case when Alice sends a cryptogram Y to Bob, then Eve, without observing Y, replaces it by some fraudulent cryptogram $\tilde{Y}$. The impersonation attack succeeds if Bob can decrypt

$Y' := \tilde{Y}$ and accepts it (i.e. believes that it comes from Alice). We denote by P_I the success probability of an impersonation attack if Alice uses an optimal strategy. There is a lower bound due to Simmons (1984) for this success probability:

Theorem 6.3. (Simmons'bound) *Denote by Z the key used. Then we have*

$$P_I \geq 2^{-I(Y;Z)}.$$

Proof: It is useful to interpret the impersonation attack as a statistical hypothesis-testing problem as follows: Denote by H_0 the null hypothesis that the cryptogram Y' received by Bob is really the cryptogram Y written by Alice, who used the key $Z = z$, hence $P(y) = P_{Y|Z}(y|z)$ for all y. As alternative, we consider the hypothesis H_1 that Y' has been formed by Eve according to the probability

$$Q(y) = P_Y(y) = \sum_z P_{Y|Z}(y|z)P_Z(z).$$

Then we have

$$D(P||Q) = E(-\log_2 \frac{P_Y(Y)}{P_{Y|Z}(Y|Z)} \quad | \quad Z = z)$$

and thus

$$\begin{aligned}
E(D(P||Q)) &= \sum_z D(P||Q)P_Z(z) \\
&= E(-\log_2 \frac{P_Y(Y)}{P_{Y|Z}(Y|Z)}) \\
&= H(Y) - H(Y|Z) \\
&= I(Y;Z).
\end{aligned} \tag{6.2}$$

But now from (6.2) and Corollary 6.1 we deduce that

$$\begin{aligned}
P_I &\geq E(\beta) \\
&\geq E(2^{-D(P||Q)}) \\
&= -E(-2^{-D(P||Q)}) \\
&\geq 2^{-E(D(P||Q))} \\
&= 2^{-I(Y;Z)}. \square
\end{aligned}$$

The significance of Simmons' bound is the following: Of course, in designing a cryptosystem where authenticity is important, one should take care that the success probability of an impersonation attack and thus Simmon's bound is as small as possible, i.e., the cryptogram should reveal a large amount of information about the key.

6.3 *Marginal Guesswork

As we have stated at the end of Section 6.1, there are also other complexity measures than entropy. Here we will present the so-called marginal guesswork, which denotes, roughly speaking, the optimal number of trials necessary to be guaranteed a certain chance α of guessing a random value in a brute-force search. It will turn out that entropy and marginal guesswork have nothing to do with each other in the sense that there is no general inequality relating them one to the other. Let X be a random variable taking values in the alphabet $\mathcal{X} = \{x_1, x_2, \ldots\}$. While entropy measures how difficult it is to determine the value of X given single queries to multiple oracles that answer questions of the type "Is $X(\omega) \in \mathcal{U}$?" for subsets $\mathcal{U} \subset \mathcal{X}$, marginal guesswork measures the difficulty of determining $X(\omega)$ with multiple queries submitted to a single oracle that answers questions "Is $X(\omega) = x$?". Let us go to a formal definition.

Assume w.l.o.g. that the probabilities $p_i := P(X = x_i)$ are sorted in non-increasing order:

$$p_1 \geq p_2 \geq \ldots \geq p_n > p_{n+1} = \ldots = 0.$$

Then, for $0 \leq \alpha \leq 1$, the α-marginal guesswork is defined as

$$w_\alpha(X) := \min\{i : \sum_{j=1}^{i} p_j \geq \alpha\}.$$

Hence, w_α measures the maximum work for determining the value of the random variable X when one wishes a probability of success of α in a brute-force search. The case $\alpha = 1$ is an exhaustive search. While in practice, the search for cipher keys is often exhaustive, the guess of passwords is rarely so (e.g., with UNIX).

If the random variable X is uniformly distributed on some subset of $\mathcal{X}$ (e.g. deterministic), then one sees at once that $H(X) \approx \log_2 w_\alpha(X)$. A similar relation holds for long random sequences with the "asymptotic Equipartition Property" (Pliam (2000), p.73). However, the two uncertainty measures "entropy" and "marginal guess work" can be completely different in the following sense:

Theorem 6.4. *For each $0 < \alpha < 1$ and every positive number N, there are finitely supported random variables X and Y such that*

$$\log_2 w_\alpha(X) > H(X) + N \tag{6.3}$$

and

$$H(Y) > \log_2 w_\alpha(Y) + N. \tag{6.4}$$

For the proof of Theorem 6.4 we need the following lemma:

Lemma 6.8. *For every $\varepsilon > 0$ there exists a finitely supported random variable X such that*

$$\sum_{i=1}^{2^{\lceil H(X) \rceil}} p_i < \varepsilon,$$

where $p_i := P(X = x_i)$ (and, w.l.o.g., $p_1 \geq p_2 \geq \ldots$).

Proof: Define the random variable $X_{j,k}$ by the sequence of probabilities

$$a^{-1}, a^{-2}, \ldots, a^{-k} \quad \text{followed by } m \text{ copies of } a^{-k},$$

where $a = 2^j$ and m is chosen so that all above probabilites sum up to 1. One observes that we must have

$$m = \frac{1 + (a-2)a^k}{a-1}.$$

Calculating the entropy gives

$$H(X_{j,k}) = \sum_{i=1}^{k} a^{-i} \log_2 a^i + \frac{1 + (a-2)a^k}{a-1} a^{-k} \log_2 a^k$$

$$= j \sum_{i=1}^{k} i a^{-i} + jk \frac{1 + (a-2)a^k}{(a-1)a^k}$$

$$= j \frac{a^{k+1} - (k+1)a + k}{(a-1)^2 a^k} + jk \frac{1 + (a-2)a^k}{(a-1)a^k}$$

$$= jk \frac{a-2}{a-1} + h_{j,k}$$

with

$$h_{j,k} = \frac{j(a^k - 1)}{a^{k-1}(a-1)^2}.$$

Now we fix a lower bound $2 < j$, hence we have $a > 4$. Then we get

$$jk \frac{a-2}{a-1} > \log_2 k$$

and hence

$$2^{\lceil H_{j,k} \rceil} \geq 2^{H_{j,k}} \geq 2^{jk(a-2)/(a-1)} > k.$$

We obtain further

$$s_{j,k} := \sum_{i=1}^{2^{\lceil H_{j,k} \rceil}} p_i$$

$$\leq \sum_{i=1}^{k} a^{-i} + (2^{\lceil H_{j,k} \rceil} - k)a^{-k}$$

$$= \frac{1}{2^j - 1} + \sigma_{j,k},$$

where

$$\sigma_{j,k} = \frac{(a-1)(2^{\lceil H_{j,k}\rceil} - k) - 1}{a^k(a-1)}.$$

If we can show that for fixed j it holds that

$$\sigma_{j,k} \to 0 \quad (k \to \infty), \tag{6.5}$$

then fixing a $j > 2$ such that $(2^j - 1)^{-1} < \varepsilon$ one can find a k such that $s_{j,k} < \varepsilon$, which finishes the proof. So it remains to prove relation (6.5). Since $h_{j,k} \to j/\alpha \ (k \to \infty)$, we may find an index $\hat{k}(j)$ such that $h_{j,k} < 1$ for all $k \ge \hat{k}(j)$. Hence for $k \ge \hat{k}(j)$ we have

$$\lceil H_{j,k}\rceil \le jk\frac{a-2}{a-1} + 2$$

and thus

$$\sigma_{j,k} \le \frac{(a-1)(4\beta^k - k) - 1}{a^k(a-1)},$$

with

$$\beta := 2^{j(a-2)/(a-1)} < a.$$

Two applications of de l'Hospital's rule yields (6.5), as desired. $\square$

Proof of Theorem 6.4: Let us first find X. We want to apply Lemma 6.8 with $\varepsilon := \alpha 2^{-N}$. We then have

$$\sum_{i=1}^{2^N 2^{\lceil H(X)\rceil}} p_i \le 2^N \sum_{i=1}^{2^{\lceil H(X)\rceil}} p_i < 2^N \varepsilon = \alpha.$$

Hence

$$w_\alpha(X) > 2^N 2^{\lceil H(X)\rceil} \ge 2^{N+H(X)},$$

which proves (6.3).

In order to prove (6.4), Y will be defined as follows: Define the probabilities $q_1 := P(Y = y_1) := \alpha$ and

$$q_i := P(Y = y_i) := (1 - \alpha)2^{-k} \quad (2 \le i \le 2^k + 1).$$

(This corresponds to a Huffman tree with one leaf of depth 1 and 2^k leaves of depth k.) One observes that $w_\alpha(Y) = 1$, while

$$H(Y) = -\alpha \log_2 \alpha - (1 - \alpha) \log_2 \frac{1-\alpha}{2^k} = (1 - \alpha)k + K(\alpha).$$

The choice

$$k > \frac{N - K(\alpha)}{1 - \alpha}$$

indeed yields (6.4). $\square$

7 Tests for (Pseudo-)Random Number Generators

In this chapter, we will present some statistical tests for (pseudo-)random number generators. As mentioned earlier, there is no "universal" test for randomness, only finitely many necessary conditions can be tested. We will orient us particularly on the list of tests that has been applied to evaluate the AES (Advanced Encryption Standard; as is known, the winner has been the RIJNDAEL algorithm ,see, e.g., Banks et al. (2000) and all the other literature on the AES, much of it available on the Internet) and the test battery suggested by Rukhin (2000b). For complete proofs, we refer to the latter paper and the literature cited therein.

7.1 The Frequency Test and Generalized Serial Test

Consider the piece

$$x := (x_{-\nu+1}, x_{-\nu+2}, \ldots, x_{N-1})$$

of a bitsequence $\{x_n\}_{n \in \mathbb{Z}}$. From this piece, one can form N overlapping ν-grams of consecutive bits. Let M_ν denote the number of pairs of repeatedly occurring ν-grams. For a fixed ν-gram

$$s := (s_{-\nu+1}, s_{-\nu+2}, \ldots, s_0)$$

it is convenient to denote the events

$$A_s(x) := \{(x_{-\nu+1}, x_{-\nu+2}, \ldots, x_0) = s\}$$

and, more generally,

$$D^m A_s(x) := \{(x_{-\nu+m+1}, x_{-\nu+m+2}, \ldots, x_m) = s\}.$$

Then one can write the test statistic as

$$M_\nu(x) = \frac{1}{2} \sum_{m,n=0; m \neq n}^{N-1} \sum_s \mathbf{1}(D^m A_s(x)) \cdot \mathbf{1}(D^n A_s(x))$$

D. Neuenschwander: Prob. and Stat. Methods in Cryptology, LNCS 3028, pp. 89-105, 2004.
© Springer-Verlag Berlin Heidelberg 2004

(where $\mathbf{1}(\ldots)$ denotes the indicator function). For $\nu = 1$ we obtain just the usual frequency test. If one defines $n_s(x)$ as the frequency of occurrence of the ν-gram s in the piece x, then one can write

$$M_\nu(x) = \sum_s \frac{1}{2} n_s(x)(n_s(x) - 1).$$

Let us determine the (asymptotic) distribution of $M_\nu(x)$ under the null hypothesis that x consists of i.i.d. unbiased random bits. Since

$$E(\mathbf{1}(D^m A_s(x))) = 2^{-\nu},$$

we obtain $E(n_s) = 2^{-\nu} N$. Let us first assume $n > m$ and $n - m < \nu$ (i.e., the "windows" are overlapping). One observes that $\mathbf{1}(D^m A_s(x)) \cdot \mathbf{1}(D^n A_s(x)) = 1$ iff the first $n - m$ bits are repeated, i.e. if x is of the form

$$(\ldots, s_{-(n-m)+1}, s_{-(n-m)+2}, \ldots, s_0, s_{-(n-m)+1}, s_{-(n-m)+2}, \ldots, s_0).$$

There are exactly 2^{n-m} ν-grams with this property, hence

$$E(\mathbf{1}(D^m A_s(x)) \cdot \mathbf{1}(D^n A_s(x))) = 2^{-(\nu+n-m)}$$

and thus

$$E(\sum_s \mathbf{1}(D^m A_s(x)) \cdot \mathbf{1}(D^n A_s(x))) = 2^{-\nu}.$$

The same formula holds in the case of non-overlapping windows. Thus

$$E(M_\nu) = N(N - 1)2^{-(\nu+1)}. \tag{7.1}$$

It is convenient to introduce the statistic

$$L_\nu := \frac{2^{\nu+1}}{N} M_\nu.$$

If we compare it with the "goodness-of-fit" statistic

$$\Psi_\nu^2 = \sum_s \frac{(n_s - 2^{-\nu} N)^2}{2^{-\nu} N}$$

$$= \frac{2^\nu}{N} \sum_s n_x(n_s - 1) + 2^\nu - N,$$

then we get

$$L_\nu = \Psi_\nu^2 - 2^\nu + N. \tag{7.2}$$

From (7.1) we obtain

$$E(L_\nu) = \frac{2^{\nu+1}}{N} E(M_\nu) = N - 1.$$

If by definition we put $\Psi_0^2 := 0$, then (7.2) hold for all $\nu \geq 0$. By a theorem of Good (1953), (1957) about the asymptotic (as $N \to \infty$) χ^2-distribution of certain statistics related to Ψ_ν^2 and some algebraic manipulations (involving first and second difference operators, applied to L_ν) one obtains the asymptotic variance

$$\text{Var}(L_\nu) = 6(2^\nu - 1) - 4\nu$$

and asymptotic covariance (for $\nu_1 < \nu_2$)

$$\text{Cov}(L_{\nu_1}, L_{\nu_2}) = 2^{\nu_1 + 1}(\nu_2 - \nu_1 + 3) - 2(\nu_1 + \nu_2 + 3).$$

Thus, for $\nu_1, \nu_2 \to \infty$, $\nu_1 < \nu_2$, the correlation coefficient has the asymptotic behavior

$$\rho(L_{\nu_1}, L_{\nu_2}) \sim \frac{3 + \nu_2 - \nu_1}{3} 2^{-(\nu_2 - \nu_1)/2}.$$

7.2 Maximum Absolute Value of Random Walk Test

If the sequence to be tested is denoted by $\{x_n\}_{n \geq 1}$, then let S_k be the k-th partial sum:

$$S_k := \sum_{j=1}^{k} x_j.$$

Take the null hypothesis of i.i.d. unbiased random bits as before. From Révész (1990), p.17 we have, for the maximal partial sum, the relation

$$P(\max_{1 \leq k \leq n} |S_k| \geq t) = 1 - \sum_{4|k| \leq (n/t)+1} P((4k-1)t \leq S_n \leq (4k+1)t)$$

$$+ \sum_{-(n/t)-3 \leq 4k \leq (n/t)-1} P((4k+1)t \leq S_n \leq (4k+3)t),$$

which can be used by noting the fact that under the null hypothesis, the statistic $(S_n + n)/2$ obeys a binomial distribution with parameters n and $p = 1/2$. However, also (even for small values of n) the following approximation is valid:

$$P(\max_{1 \leq k \leq n} S_k \leq \sqrt{n}z) \overset{n \to \infty}{\to} \frac{4}{\pi} \sum_{j=0}^{\infty} \frac{(-1)^j}{2j+1} \exp(-\frac{((2j+1)\pi)^2}{8z^2})$$

$$\overset{z \to \infty}{\sim} 1 - \frac{4}{3\sqrt{2\pi}z} \exp(-\frac{9z^2}{2})$$

(see Rukhin (2000b)).

7.3 Number of Visits of Random Walk Test

An excursion of the partial sum process mentioned in the previous section is a sequence of indexes

$$(i, i+1, \ldots, \ell): \quad S_{i-1} = S_{\ell+1} = 0, \quad S_k \neq 0 \ (k = i, i+1, \ldots, \ell).$$

Let J be the number of excursions. The null hypothesis of i.i.d. unbiased random bits should be rejected if the value

$$P(J < J(obs)) \approx \sqrt{\frac{2}{\pi}} \int\limits_{0}^{J(obs)/\sqrt{n}} e^{-u^2/2} du$$

is too small (where $J(obs)$ means the observed value of J). If this is not the case, then the statistic $\xi(x)$, defined as the number of visits to $x(\neq 0)$ in one excursion should be calculated. Its distribution under the null hypothesis is known to be as follows:

Proposition 7.1.

$$P(\xi(x) = 0) = 1 - \frac{1}{2|x|}$$

and

$$P(\xi(x) = k) = \frac{1}{4x^2}(1 - \frac{1}{2|x|})^{k-1} \quad (k \geq 1).$$

Proof: We first assume that the individual random bits s_j are i.i.d. *biased* random bits, i.e., $P(S_j = 1) = p$, $P(S_j = 0) = q < p$. In a finite excursion it holds that $\xi(x) = k \geq 1$ iff the random walk $\{S_k\}_{k \geq 0}$ attains the level x, then visits x exactly $k - 1$ times before it finally returns to zero. Hence the shifted random walk $\{S_k - x\}_{k \geq 0}$ never attains the value $-x$ during its first $k - 1$ excursions. Denote $\rho := \min\{k \geq 1 : S_k = 0\}$. Since the individual excursions are independent, one obtains

$$P(\xi(x) = k, \rho < \infty) = P(\xi(x) > 0)(P(\xi(-x) = 0, \rho < \infty))^{k-1} P(\xi(-x) > 0).$$
(7.3)

Assume first that $x > 0$. Let π be the probability that $\{S_k\}_{k \geq 0}$ visits $x - 1$ before -1. We get

$$P(\xi(x) > 0) = P(S_1 = 1)\pi.$$

The probability π can be calculated by Pascal's ruin problem (see, e.g., Feller (1968), Section XIV.2, (2.5)), with which we obtain

$$P(\xi(x) > 0) = \frac{q - p}{(q/p)^x - 1}.$$

Similarly,

$$P(\xi(-x) > 0) = \frac{p - q}{(p/q)^x - 1},$$

hence together

$$P(\xi(x) > 0) = |\frac{p - q}{1 - (q/p)^x}|$$

for all x; letting $p \to 1/2$ proves the first assertion of the proposition. On the other hand, if we put $I := 1\{x > 0\}$, then the above gives

$$P(\xi(x) = 0, \rho < \infty) = 1 - |\frac{p - q}{1 - (q/p)^x}| - (1 - I)(p - q), \qquad (7.4)$$

since

$$P(\xi(x) = 0, \rho = \infty) = (1 - I)P(\rho = \infty) = (1 - I)(p - q). \qquad (7.5)$$

Substituting this in (7.3) yields

$$P(\xi(x) = k, \rho < \infty)$$
$$= (pq)^x (\frac{p - q}{p^x - q^x})^2 (1 - \frac{p - q}{|(p/q)^x - 1|} - I(p - q))^{k-1}. \qquad (7.6)$$

If one has an infinite excursion, then $\xi(x) = k$ iff $\{S_k\}_{k\geq0}$ visits x, then returns to x exactly $k - 1$ times (but does not return to 0) without visiting x and 0 afterwards. This is possible only if $x > 0$, since otherwise $\{S_k\}_{k\geq0}$ must attain 0 again. So if $I = 0$, then $P(\xi(x) = k, \rho = \infty) = 0$. If we replace x by $-x$ in (7.4), we obtain

$$P(\xi(x) = k, \rho = \infty) = P(\xi(x) > 0)P(\xi(-x) = 0, \rho < \infty)^{k-1}P(\rho = \infty)$$
$$= \frac{I(p - q)^2}{1 - (\frac{q}{p})^x}(1 - \frac{p - q}{(\frac{p}{q})^x - 1} - (p - q))^{k-1}.$$

Adding this to (7.6), and letting $p \to 1/2$, then the second assertion of the proposition follows. $\square$

Now one can test the observed values $\xi(x)(obs)$ against the theoretical ones by a chi-square test.

7.4 Run Tests

There are different definitions of runs in bitsequences. In the sequel, we will use the definition due to Feller: A bitsequence of length n contains as many 0-runs of length m as there are non-overlapping uninterrupted blocks containing exactly m zeroes each. The 1-runs are defined similarly. Let

$$\mu := 2^{m+1} - 2$$

and

$$\sigma^2 := 2^{2(m+1)} - (2m+1)2^{m+1} - 2$$

and define $W(m, n)$ to be the number of runs of length m. By the limit theorem (see Rukhin (2000b), (5))

$$P(\frac{(W(m, n) - (n/\mu))\mu^{3/2}}{\sqrt{n}\sigma} < z) \to \Phi(z) \quad (n \to \infty)$$

($\Phi(z)$ denoting the standard normal distribution function), it holds that for

$$z(obs) = \sqrt{\mu}(\mu(W(m, n)(obs)) - n)/(\sigma\sqrt{n})$$

the asymptotic p-value (as $n \to \infty$) is given by

$$p = 2(1 - \Phi(|z(obs)|)).$$

Now we consider the case where also $m \to \infty$, like

$$\frac{n}{2^m} \to \lambda > 0.$$

In this case, we have that $W(m, n)$ tends weakly to a Poisson distribution with parameter λ (see Rukhin (2000b), (7), Barbour et al. (1992), Section 8.4). On the other hand, if one denotes by $\tilde{W}(m, n)$ the number of overlapping runs of length m, then $\tilde{W}(m, n)$ tends weakly to the so-called Polya-Aeppli distribution with Laplace transform (moment-generating function)

$$E(e^{t\tilde{W}}) = \exp(\frac{\lambda(e^t - 1)}{1 - e^t/2})$$

(see Rukhin (2000b), (8)). This turns out to be a compound Poisson distribution, i.e. it corresponds to a random variable U with law

$$P(U = u) = e^{-\lambda/2}2^{-u}\sum_{\ell=1}^{u}\binom{u-1}{\ell-1}\frac{\lambda^\ell}{2^\ell\ell!} \quad (u \geq 1)$$

(see Rukhin (2000b), 5.1). This latter expression can also be written in terms of the confluent hypergeometric function $_1F_1$:

$$P(U = u) = \frac{\lambda e^{-\lambda}}{2^{u+1}}{_1F_1}(u + 1, 2, \lambda/2) \quad (u \geq 1).$$

To use this result for a test, one partitions the observed bitsequence into N substrings and the empirical frequencies within each such substring are conjoined by the χ^2-statistic.

To test randomness, the Longest Run Test is also appropriate. Let ν denote the length of the longest run in a sequence of length $n = MN$ (N blocks of size M). If in the block (of size M) one has m ones and if we put $u := \min\{M - m + 1, \lfloor m/(k+1) \rfloor\}$, then

$$P(\nu \le k) = 2^{-M} \sum_{r=0}^{M} \binom{M}{r} P(\nu \le k \mid r),$$

$$P(\nu \le k \mid r) = \binom{M}{r}^{-1} \sum_{j=0}^{u} (-1)^j \binom{M - r + 1}{j} \binom{M - j(k+1)}{M - r}$$

(see Rukhin (2000b), p.117, Barton, David (1962)).

7.5 Tests on Frequencies of Patterns

Let

$$s = (s_1, s_2, \ldots, s_m)$$

be a nonperiodic pattern (template) of length $m = \log_2(n/\lambda)$. Let $W(m, n)$ be the number of occurrences of s in a bitstring of length n. Take the usual null hypothesis that the bistring consists of i.i.d. unbiased random bits. By interpreting $W(m, n)$ suitably as a sum of indicator functions that the observed substring of length m coincides with s, one obtains

$$E(W) = (n - m + 1)2^{-m}.$$

If $2^{-m}n \to \lambda$, then $W(m, n)$ has asymptotically a Poisson distribution with parameter λ (Rukhin (2000b), 5.1, Barbour et al. (1992), Section 8.4). If only $n \to \infty$ and m remains fixed, then the law of $W(m, n)$ (under suitable normalization) tends to a standard normal distribution (see also Rukhin (2000b), 5.1).

7.6 Tests Based on Missing Words

We first make a preparation on correlation polynomials. Denote, as usually, by $\{x_n\}_{n \ge 1}$ a sequence of i.i.d. unbiased random bits. Let $s, t \in \mathbb{B}^m$ be aperiodic templates (patterns) of length m. Then the cprrelation polynomial is defined as

$$C_{s,t}(z) := \sum_{k=1}^{m} \delta_{(s_{m-k+1},\ldots,s_m),(t_1,\ldots,t_k)} 2^{-(m-k)} z^{k-1}$$

(where $\delta_{\cdot,\cdot}$ denotes the Kronecker symbol: $\delta_{s,t} := 1$ if $s = t$ and zero else). The autocprrelation polynomial is defined as

$$A_s(z) := C_{s,s}(z).$$

By the aperiodicity of s, we have $A_s(z) = z^{m-1}$. The autocprrelation matrix is defined as the (2×2)-matrix

$$\mathcal{A}(z) = \begin{pmatrix} A_s(z) & C_{s,t}(z) \\ C_{t,s}(z) & A_t(z) \end{pmatrix}.$$

Let $\pi_s(n)$ be the probability that the template s is missing in $\{x_1, x_2, \ldots, x_n\}$. Then by a theorem due to Guibas and Odlyzko (1981) we have that

$$F_s(z) := \sum_{n=m}^{\infty} \frac{\pi_s(n)}{z^n} \tag{7.7}$$

$$= \frac{z A_s(z)}{(z-1)A_s(z) + 2^{-m}}. \tag{7.8}$$

Let us write the above expression in the form

$$\sum_{n=m}^{\infty} \frac{\pi_s(n)}{z^n} = \frac{z A_s(z)}{P(z)}, \tag{7.9}$$

where

$$P(z) = \prod_{j=1}^{m} (z - z_j)$$

is a polynomial of degree m with leading coefficient 1. On the other hand, we observe that

$$F_s(z) = \sum_{k=1}^{m} \delta_{(s_{m-k}, \ldots, s_m),(t_1, \ldots, t_k)} 2^{-(m-k)}$$

$$\cdot \sum_{n=0}^{\infty} z^{-(n+m-k)} \sum_{k_1 + \ldots + k_m = n} \prod_{j=1}^{m} z_j^{k_j}. \tag{7.10}$$

By comparing coefficients in (7.7) and (7.10) we obtain finally

$$\pi_s(n) = \sum_{k_1 + \ldots + k_m = n} \prod_{j=1}^{m} z_j^{k_j}.$$

Let $n, m \to \infty$ such that

$$n2^{-m} \to a > 0.$$

One can show that in this case the asymptotic behavior of $\pi_s(n)$ is of the form

$$\pi_s(n) \sim e^{-a}(1 - (2m-1)a2^{-(m+1)} + (m-1)2^{-m}) \tag{7.11}$$

(see Rukhin (2000b), p.120). Let X be the number of missing templates of length m in $\{x_1, x_2, \ldots, x_n\}$. Then one obtains the following asymptotic behavior of the expectation of X:

$$E(X) = e^{-a}2^m + e^{-a}(m - 1 - a/2) + O(1)$$

(Rukhin (2000b), (18)). By a similar, but of course more cumbersome procedure, one can also derive the variance of X:

$$\text{Var}(X) = \sum_s \pi_s(n)(1 - \pi_s(n)) + \sum_{s \neq t}(\pi_{s,t}(n) - \pi_s(n)\pi_t(n)),$$

where $\pi_{s,t}(n)$ denotes the probability that templates s and t are missing in $\{x_1, x_2, \ldots, x_n\}$ (Rukhin (2000b), (19)). These probabilities can be determined from the equation (by comparison of coefficients)

$$\sum_{n=1}^{\infty} \frac{\pi_{s,t}(n)}{z^n}$$

$$= z \cdot \det(\mathcal{A}(z))((z - 1)\det(\mathcal{A}(z)) + 2^{-m}(A_s(z) + A_t(z) - C_{s,t}(z) - C_{t,s}(z)))^{-1}.$$

The asymptotic behavior of the probabilities $\pi_{s,t}(n)$ is of the following type: If $C_{s,t}(z) = C_{t,s}(z) = 0$, then

$$\pi_{s,t}(n) = e^{-2a}(1 + (m - 1 - (2m - 1)a)2^{-(m-1)}) + O(2^{-2m}).$$

If $C_{s,t}(z)$ is of degree $m - 1 - u$ and $C_{t,s}(z) \equiv 0$, then

$$\pi_{s,t}(n) = e^{-2a}(1 + a2^{-u}) + O(2^{-\min\{m,2u\}}).$$

It turns out that the main contribution is given by these two types of pairs (s, t).

7.7 Approximate Entropy Test

A more general notion than entropy is the so-called ϕ-entropy. Assume $\phi : [0, 1] \to \mathbb{R}$ is a convex C^2-function with $\phi(1) = 0$. Then the ϕ-entropy of a random variable X having discrete distribution μ with atoms $p_1, p_2, \ldots, p_M > 0$ (at some places) is defined as

$$\sum_{j=1}^{M} p_j \phi(p_j).$$

The ϕ-entropy with $\phi = -\log_2$ is just the usual entropy. If $M = 2^m$ and the probability law μ is just the distribution of all templates s of length m, then one defines the ϕ-uncertainty as

$$\sum_s \nu_s \phi(\nu_s)$$

(ν_s denoting the relative frequency of the template s in the augmented (circular) extension of the original bitstring). In order to get a limiting distribution for this statistic (under the usual null hypothesis), we normalize it as follows:

$$\Phi^{(m)} := a_m \sum_s \nu_s \phi(\nu_s),$$

with

$$a_m := \frac{2^m}{\phi'(2^{-m})2^{-(m+1)}\phi''(2^{-m})}.$$

Now the "approximate ϕ-entropy of order m" is defined as

$$AH(m) = \Phi^{(m)} - \Phi^{(m+1)}$$

(see Rukhin (2000a,b)). The limit distribution (as $n \to \infty$, n denoting again the length of the bitstring to be tested), after centering, is as follows:

$$\mathcal{L}(nAH(m) - a_m\phi(2^{-m}) + a_{m+1}\phi(2^{-m-1})) \xrightarrow{w} \chi^2(2^{m+1} - 2^m)$$

(see Rukhin (2000b), p.123, Mitra, Rao (1971), Theorem 9.2.2). If we define the classical Pearson χ^2-statistics as Ψ_m^2, i.e.,

$$\Psi_m^2 = \sum_s \frac{(n\nu_s - n2^{-m})^2}{n2^{-m}},$$

then its relation to $\Phi^{(m)}$ is that

$$\Phi^{(m)} = 1 + \frac{\Psi_m^2}{n},$$

and thus for the difference sequence

$$\Delta\Psi_m^2 := \Psi_m^2 - \Psi_{m-1}^2 = nAH(m-1),$$

whose law converges weakly, as $n \to \infty$, to $\chi^2(2^{m-1})$. Also, it can be shown that the laws of the second differences

$$\Delta^2\Psi_m^2 := (\Psi_m^2 - \Psi_{m-1}^2) - (\Psi_{m-1}^2 - \Psi_{m-2}^2)$$

converge weakly (as $n \to \infty$) to $\chi^2(2^{m-2})$. The Pearson χ^2-statistic as mentioned here corresponds to the choice of $\phi(u) = u$ (see Rukhin (2000b), p.124; see also Billingsley (1956)).

7.8 The Ziv-Lempel Complexity Test

Here, the test statistic $W(n)$ is defined (recursively) as the number of words that arise if the bitsequence of length n is parsed into consecutive disjoint words such that the next word is the shortest template not seen before. Loosely speaking, this is a test of compressibility of the source. The statistic $W(n)$ behaves as follows (under the null hypothesis): First,

$$\frac{E(W(n))}{n/\log_2 n} \to 1 \quad (n \to \infty)$$

(a result due to Aldous, Shields (1988); see Rukhin (2000b)). Furthermore, there is a constant $\sigma > 0$ such that distribtution of

$$\frac{W(n) - E(W(n))}{\sigma W(n)}$$

tends weakly to a standard normal law. The value of σ has the property

$$\sigma^2 W(n) \overset{w}{\sim} \frac{n(C + h(\log_2 n))}{\log_2^3 n} \quad (n \to \infty),$$

where $C \approx 0.26600$ is a constant and h is a random slowly varying continuous function with zero mean and $|h(.)| < 10^{-6}$ (Kirschenhofer et al. (1994), see Rukhin (2000b). p.125).

7.9 Maurer's "Universal Test"

Maurer calls his test "universal" because "it can detect any significant deviation of a device's output statistics from the statistics of a truly random bit source when the device can be modeled as an ergodic stationary source with finite memory but arbitrary (unknown) state transition probabilities" (Maurer (1992)). The statistic of Maurer's test is closely related to the entropy per bit of the source, which is "the correct quality measure for a secret-key source in a cryptographic application" (Maurer (1992)). Perhaps, in our context, the word "universal" should be written in quotation marks; as we have stated before, there are no practically implementable universal (in the literal sense of the word as used in our text) tests of randomness. As the previously discussed Ziv-Lempel Complexity Test, the statistic of Maurer's test measures the compressibility of the sequence. If the bitstream is significantly compressible, then it should be considered as non-random. Maurer (1992) rather discourages using the Ziv-Lempel complexity test. On the other hand, the disadvantage of Maurer's test is that one must have an x of length n where n is of the order $10 \cdot 2^L + 1000 \cdot 2^L$ with $6 \le L \le 16$. The first $Q = 10 \cdot 2^L$ blocks of L bits serve as initialization blocks, whereas the last $K := \lfloor n/L \rfloor - Q$ blocks of length L are the test blocks. The size of Q makes sure that with high probability, all L-bit strings occur in the initialization blocks. Now Maurer's test statistic is the following:

$$f_n := \frac{1}{K} \sum_{i=Q+1}^{Q+K} \log_2 t_i$$

where t_i denotes the number of indices since the previous occurrence of the i-th template. In other words, the test consists of looking back through the entire sequence while inspecting the test segment of L-bit blocks, checking for the nearest previous L-bit template match, and recording the distance (in number of blocks) to that previous match. One finds that under the null hypothesis of i.i.d. unbiased bits, the expectation of f_n is given as

$$E(f_n) = 2^{-L} \sum_{i=1}^{\infty} (1 - 2^{-L})^{i-1} \log_2 i.$$

The variance can be approximately calculated as follows:

$$\text{Var}(f_n) = \frac{c(L,K)}{K} \text{Var}(\log_2 G),$$

where G denotes a geometrically distributed random variable (with parameter $1 - 2^{-L}$) and $c(L,K)$ has the approximate value

$$c(L,K) \approx 0.7 - \frac{0.8}{L} + (1.6 + \frac{12.8}{L})K^{-4/L}.$$

However, Coron and Naccache (1999), who confirmed this approximation, warn that "the inaccuracy due to this approximation can make the test to be 2.67 times more permissive than what is theoretically admitted". So it is also reasonable to test the hypothesis of randomness by verifying the normality of the observed values $f_n(obs)$ by the t-test, where the variance is unknown. For running this t-test, one should partition the observed sequence in a number of, say, $r \leq 20$ substrings, for every one of which one runs the test statistic, then calculates the sample variance, and finally determines the p-value from the t-distribution with $r - 1$ degrees of freedom (Rukhin (2000b)).

7.10 Rank of Random Matrices Test

Let R be the rank of an $M \times Q$-random matrix with entries in $I\!B$. The possible values of R are $0, 1, \ldots, m := \min\{M, Q\}$. By a calculation due to Kovalenko (1972), the random variable R obeys (under the null hypothesis) the following distribution:

$$P(R = r) = 2^{r(Q+M-r)-MQ} \prod_{i=0}^{r-1} \frac{(1 - 2^{i-Q})(1 - 2^{i-M})}{1 - 2^{i-r}}.$$

Take $M = Q \geq 10$. Then we may approximate

$$P(R = M) \approx \prod_{j=1}^{\infty}(1 - 2^{-j}) \approx 0.2888\ldots,$$

$$P(R = M - 1) \approx 2P(R = M) \approx 0.5776\ldots,$$

$$P(R = M - 2) \approx \frac{4}{9}P(R = M) \approx 0.1284\ldots,$$

whereas $P(R = r) \leq 0.005$ for $R \notin \{M - 2, M - 1, M\}$. Let $N \approx n/M^2$ (where n is the length of the observed bitstring). So N can be interpreted as the new "sample size", i.e., we can form N random (square) matrices with the observed input sequence. We calculate their ranks $R_1, R_2, \ldots, R_N$ and determine the frequencies F_M, F_{M-1}, F_{M-2} of the rank values $M, M-1, M-2$ resp. among the $R_1, \ldots, R_N$. Then we apply the chi-square test: The test statistic

$$\chi^2 = \frac{(F_M - 0.2888N)^2}{0.2888N} + \frac{(F_{M-1} - 0.5776N)^2}{0.5776N}$$
$$+ \frac{(N - F_M - F_{M-1} - 0.1336N)^2}{0.1336N}$$

has, under the null hypothesis of independent unbiased bits, a chi-square distribution with 2 degrees of freedom. The p-value is

$$p = \exp(-\chi^2(obs)/2).$$

7.11 Linear Complexity Test

The linear complexity of a finite bitsequence $x = \{x_i\}_{0 \leq i \leq n}$ of length $n + 1$ is defined as the length of the shortest LFSR over the field $\mathbb{B}$ that generates x. We refer to Section 5.1 for more information about LFSR, Such a shortest LFSR can be determined by the famous Berlekamp-Massey algorithm, which we will present in the following. (A generalization of the Berlekamp-Massey algorithm to residue rings was given by Reeds, Sloane (1985). We stress that high linear complexity is by far not sufficient for a sequence to be considered as "random". E.g., the sequence $0, 0, \ldots, 0, 1$ has maximal linear complexity, but is very "regular"! Let L be the length of a LFSR. We can say that the LFSR $(c_0, c_1, \ldots, c_L) \in \mathbb{B}^{L+1}$ (where $c_L = 1$) generates the sequence x if

$$\sum_{i=0}^{L} c_i x_{j+i} = 0 \quad (0 \leq j \leq n - L).$$

For the Berlekamp-Massey algorithm it is useful to work with polynomials, since polynomial rings have "nice" algebraic properties. So we call the polynomial

$$c_{(n)}(z) := \sum_{i=0}^{L} c_i z^i$$

(which is of degree L) the characterisitc polynomial (or recursion polynomial) of the LFSR (see Section 5.1). In order to make explicit the dependence of the coefficients of n, let us write $c_i =: c_{n,i}$ in the above formula. The Berlekamp-Massey algorithm is recursive and runs as follows: Suppose we have found a characteristic polynomial $c_{(k-1)}(z)$ of degree L_{k-1} that generates the partial sequence $x^{(k)} = \{x_i\}_{0 \leq i \leq k-1}$ of length k. We want to find a characteristic polynomial $c_{(k)}(z)$ that generates $x^{(k+1)}$.The length L_k of this new LFSR will be the degree of $c_{(k)}(z)$. So we have

$$\sum_{i=0}^{L_{k-1}} c_{k-1,i} x_{j+i} = 0 \quad (0 \leq j \leq k-1-L_{k-1}).$$

On the other hand,

$$\sum_{i=0}^{L_{k-1}} c_{k-1,i} x_{k-L_{k-1}+i} = \delta_k$$

for certain $\delta_k \in \mathbb{B}$. Now the recursion step is as follows: If $\delta_k = 0$, then $c_{(k-1)}(z)$ also generates x_k and we can take

$$c_{(k)}(z) = c_{(k-1)}(z)$$

and hence

$$L_k := L_{k-1}.$$

The more difficult case is when $\delta_k = 1$. Let m be such that

$$L_{m-1} < L_m = L_{k-1}$$

(i.e., the length of the LFSR before the last jump of the length in the recursion). Then we have

$$L_k = \max\{L_{k-1}, k+1-L_{k-1}\} \tag{7.12}$$

and one possible choice for a characteristic polynomial $c_{(k)}(z)$ for $x^{(k+1)}$ is the following:

$$c_{(k)}(z) := z^{L_k - L_{k-1}} c_{(k-1)}(z) - z^{L_k - (k-m+L_{m-1})} c_{(m-1)}(z). \tag{7.13}$$

(Remark: In general, the characteristic polynomials of minimal degree are not uniquely determined. One can find all of them, but this is not of significance here, since we are only interested in the length of the shortest LFSR.)
Proof of the Berlekamp-Massey Algorithm: 1. First we prove the inequality

$$L_k \geq \max\{L_{k-1}, k+1-L_{k-1}\}. \tag{7.14}$$

The relation $L_k \geq L_{k-1}$ being trivial, it remains to prove

$$L_k \geq k + 1 - L_{k-1}. \tag{7.15}$$

Assume that $c_{(k-1)}(z)$ generates $x^{(k)}$, but not $x^{(k+1)}$. Then it is not difficult to show (by some elementary manipulations with Laurent series) that there exist polynomials $p(z)$ of degree $< L_{k-1}$ and $\overline{p}(z)$ of degree $< L_k$ such that the following Laurent series expansions hold:

$$z \frac{p(z)}{c_{(k-1)}(z)} = \sum_{i=0}^{k-1} x_i z^{-i} + y_k z^{-k} + \dots$$

and

$$z \frac{\overline{p}(z)}{c_{(k)}(z)} = \sum_{i=0}^{k} x_i z^{-i} + \dots$$

such that $x_k \neq y_k$. Hence it follows that

$$p(z)c_{(k)}(z) - \overline{p}(z)c_{(k-1)}(z) = c_{(k-1)}(z)c_{(k)}(z)((y_k - x_k)z^{-(k+1)} + \dots).$$

Since $x_k \neq y_k$, it follows that $L_{k-1} + L_k - (k+1) \geq 0$, which proves (7.15).
2. Now we are ready to prove (7.12) by induction. The induction begins at the first jump of the sequence $\{L_i\}$., i.e., for

$$x^{(\ell+1)} = (0, 0, \dots, 0, 1) \in \mathbb{B}^{\ell+1}.$$

Here we have $L_{\ell-1} = 0$ and $L_\ell = \ell + 1 = \max\{0, \ell + 1 - 0\}$, hence (7.12) (for the beginning of the induction) is fulfilled. For the induction step, assume that (7.12) is valid for $k = m$, i.e.,

$$L_m = L_{k-1} = \max\{L_{m-1}, m + 1 - L_{m-1}\}.$$

Since $L_m > L_{m-1}$ it follows that $L_{k-1} = L_m = m + 1 - L_{m-1}$, hence

$$k - m + L_{m-1} = k + 1 - L_{k-1}. \tag{7.16}$$

Now since $c_{/k)}(z)$ is of degree L_k by definition, it suffices to show that it indeed generates $x^{(k+1)}$, for in this case it is (by 1.) a generating LFSR of minimal length. Put

$$x^{(k+1)}(z) := \sum_{i=0}^{k} x_i z^{-i}.$$

Write the Laurent series expansions

$$c_{(m-1)}(z)x^{(k+1)}(z) =: \sum_{i=0}^{\infty} \alpha_i z^{-i}$$

and

$$c_{(k-1)}(z)x^{(k+1)}(z) =: \sum_{i=0}^{\infty} \beta_i z^{-i}.$$

Then one can show that

$$\alpha_i = 0 \quad (0 \le i \le m - 1 - L_{m-1}),$$

$$\alpha_{m-L_{m-1}} = 1,$$

$$\beta_i = 0 \quad (0 \le i \le k - 1 - L_{k-1}),$$

$$\beta_{k-L_{k-1}} = 1.$$

If we develop, furthermore,

$$z^{L_k - L_{k-1}} c_{(k-1)}(z)x^{(k+1)}(z) =: \sum_i \gamma_i z^{-i}$$

and

$$z^{L_k - (k-m+L_{m-1})} c_{(m-1)}(z)x^{(k+1)}(z) =: \sum_i \nu_i z^{-i},$$

then we get the facts that

$$\gamma_{i-L_k} = 0 \quad (L_{k-1} \le i \le k - 1),$$

$$\gamma_{k-L_k} = 1,$$

$$\nu_{i-L_k} = 0 \quad (k - m + L_{m-1} \le i \le k - 1),$$

$$\gamma_{k-L_k} = 1.$$

Hence we obtain

$$\nu_{j-L_k} = 0 \quad (k + 1 - L_{k-1} \le j \le k - 1)$$

and

$$\nu_{k-L_k} = 1.$$

Hence in the power series expansion of the product

$$c_{(k)}(z)x^{(k+1)}(z) =: \sum_{i=0}^{\infty} \mu_i z^i$$

we have $\mu_j = 0$ for $0 \le j \le k - L_k$, which means that $c_{(k)}(z)$ indeed generates the sequence $x^{(k+1)}$. $\square$

Rueppel (1986) gives the distribution of L_n under the null hypothesis of a genuine random sequence: From (7.12) it follows that $N_n(L)$, which means the number of bitsequences of length n and linear complexity L, is given by

$$N_n(L) = \begin{cases} 2N_{n-1}(L) + N_{n-1}(n - L) & : \quad n \ge l > n/2 \\ 2N_{n-1}(L) & : \quad L = n/2 \\ N_{n-1}(L) & : \quad n/2 > L \ge 0. \end{cases}$$

From this, the following corollary can be proved by induction on n:

Corollary 7.1. $P(L_n = 0) = 2^{-n}$ and

$$P(L_n = L) = \frac{2^{\min\{2n-2L, 2L-1\}}}{2^n} \quad (1 \le L \le n).$$

By standard analytic calculations, the expectation, variance, and (more generally) the generating function can be evaluated:

$$E(L_n) = \frac{n}{2} + \frac{9 + (-1)^{n+1}}{36} - \frac{1}{2^n}\left(\frac{n}{3} + \frac{2}{9}\right),$$

$$\begin{aligned}
\mathrm{Var}(L_n) &= \frac{86}{81} - 2^{-n}\frac{14 - ((1 - (-1)^n)/2)}{27}n + \frac{81 + (-1)^n}{81} \\
&\quad - 2^{-2n}\left(\frac{1}{9}n^2 + \frac{4}{27}n + \frac{4}{81}\right) \\
&\to \frac{86}{81} \quad (n \to \infty)
\end{aligned}$$

(see Rueppel (1986)) and

$$\begin{aligned}
E(e^{tL_n}) &= \frac{1}{2^n}\left(\frac{1}{2} + \frac{e^{(\lfloor n/2\rfloor + 1)(t + 2\log 2)} - 1}{2(4e^t - 1)}\right. \\
&\quad \left. + e^{tn}\frac{e^{(n - \lfloor n/2\rfloor)(-t + 2\log 2)} - 1}{4e^{-t} - 1}\right).
\end{aligned}$$

For the limiting distribution of L_n, one finds

$$\mathcal{L}\left((-1)^n\left(L_n - \frac{n}{2} - \frac{4 + (1 - (-1)^n)/2}{18}\right) + \frac{2}{9}\right) \overset{w}{\to} \mathcal{L}(T),$$

where $P(T = 0) = \frac{1}{2}$,

$$P(T = k) = 2^{-2k} \quad (k \ge 1),$$

and

$$P(T = k) = 2^{-2|k|+1} \quad (k \le -1)$$

(see Rukhin (2000b)).

8 Diffie-Hellman Key Exchange

8.1 The Diffie-Hellman System

Here, we will look at another public-key system, namely the Diffie-Hellman key distribution algorithm (Diffie, Hellman (1976)). It works as follows: Let α be a fixed non-multiple of a prime p. First, Alice chooses her private key $x_A \in \mathbb{Z}_{p-1}$. She determines her public key y_A by

$$y_A = \alpha^{x_A} \in \mathbb{Z}_p^*.$$

The same is done by Bob. Now, if Alice and Bob want to generate a secret Diffie-Hellman key, Alice requests Bob's public key y_B from the directory where it is published and generates the Diffie-Hellman key

$$y_B^{x_A} = \alpha^{x_A x_B}.$$

Bob does the same mutatis mutandis and gets

$$y_A^{x_B} = \alpha^{x_B x_A},$$

which turns out to be the same as Alice's Diffie-Hellman key! The security of the Diffie-Hellman procedure rests on the discrete logarithm problem. It is generally believed that solving congruential equations $a^z = b(mod.n)$ with respect to z (with a, b, n given) is computationally difficult, in a certain sense perhaps even harder than factoring integers. However, here also, it has not been proved that solving the discrete logarithm problem is really necessary for breaking the Diffie-Hellman system. Other cryptosystems based on the difficulty of the discrete logarithm problem are the ElGamal and the Massey-Omura system (see Beutelspacher (1993), p.141). The following considerations are based on Massey, Waldvogel (1993). See this paper for further details.

8.2 Distribution of Diffie-Hellman Keys

In this section, we want to give some information about the probability distribution of the keys in the Diffie-Hellman system. We start with the general

D. Neuenschwander: Prob. and Stat. Methods in Cryptology, LNCS 3028, pp. 107-113, 2004.
© Springer-Verlag Berlin Heidelberg 2004

result and then we will show how the modulus parameter p should be chosen to get a "good" distribution (i.e., an "almost equidistribution") of the Diffie-Hellman keys.

Theorem 8.1. *Let p be a prime and denote by $p - 1 = \prod_{i=1}^{K} p_i^{e_i}$ the prime factorization of $p - 1$. Furthermore, suppose $t \in \mathbb{Z}_{p-1}$ and α a generator of $\mathbb{Z}_p^*$. Let*

$$m(\alpha^t) = \prod_{i=1}^{K} p_i^{\tilde{e}_i}$$

$(0 \leq \tilde{e}_i \leq e_i)$ be the multiplicative order of α^t. Put

$$R(t) := |\{(x_A, x_B) \in \mathbb{Z}_{p-1} \times \mathbb{Z}_{p-1} : \alpha^{x_A x_B} = \alpha^t\}|.$$

Then we have

$$R(t) = \prod_{i=1}^{K} p_i^{e_i - 1}((p_i - 1)(e_i - \tilde{e}_i + 1) + \delta(\tilde{e}_i))$$

(where $\delta(0) := 1$ and $\delta(e) := 0$ $(e \neq 0)$).

For the proof of Theorem 8.1 we need some algebraic and combinatoric preparations. We start with the following easy lemma (see, e.g., Hardy, Wright (1960), Theorem 57, Massey, Waldvogel (1993), Lemma 1):

Lemma 8.1. *The equation*

$$x_A x_B = t$$

over $\mathbb{Z}_{p-1}$ has solutions for x_B iff (over $\mathbb{Z}$) we have that $\gcd(x_A, p - 1)|t$; moreover, in the latter case, the number of solutions for x_B is $\gcd(x_A, p - 1)$.

So one can write

$$R(t) = \sum_{x_A \in S(t)} \gcd(x_A, p - 1), \qquad (8.1)$$

where

$$S(t) := \{u \in \mathbb{Z}_{p-1} : \gcd(u, p - 1)|t\}.$$

Lemma 8.2. *Assume $t \in \mathbb{Z}_{p-1}$. Then for all $u \in \mathbb{Z}$ we have that $\gcd(u, p - 1)|t$ iff $\gcd(u, p - 1)|\gcd(t, p - 1)$.*

Proof: If $\gcd(u, p - 1)|t$, then (since also $\gcd(u, p - 1)|p - 1$) it follows that $\gcd(u, p - 1)|\gcd(t, p - 1)$. On the other hand, if $\gcd(u, p - 1)|\gcd(t, p - 1)$, then (since $\gcd(t, p - 1)|t$), it follows that $\gcd(u, p - 1)|t$. $\square$

Hence

$$S(t) = \{u \in \mathbb{Z}_{p-1} : \gcd(u, p - 1)|\gcd(t, p - 1)\}. \qquad (8.2)$$

By a fact from elementary algebra (see Lidl, Niederreiter (1986), Theorem 1.15ii)), we have

$$m(\alpha^t) = \frac{p-1}{\gcd(t, p-1)}. \tag{8.3}$$

If we substitute this into (8.2), we obtain

$$S(t) = \{u \in \mathbb{Z}_{p-1} : \gcd(u, p-1) | \frac{p-1}{m(\alpha^t)}\}. \tag{8.4}$$

Decompose $p - 1$ into prime factors: $p - 1 = \prod_{i=1}^{K} p_i^{e_i}$ (with $e_i \geq 1$ and p_i distinct prime factors). Lagrange's Theorem, applied to $\alpha^t \in \mathbb{Z}_p^*$ yields

$$m(\alpha^t) = \prod_{i=1}^{K} p_i^{\tilde{e}_i},$$

with $0 \leq \tilde{e}_i \leq e_i$. So if we substitute

$$\frac{p-1}{m(\alpha^t)} = \prod_{i=1}^{K} p_i^{e_i - \tilde{e}_i}$$

into (8.4), we obtain

$$S(t) = \{u \in \mathbb{Z}_{p-1} : \gcd(u, p-1) | \prod_{i=1}^{K} p_i^{e_i - \tilde{e}_i}\}.$$

If we consider the prime factorization

$$\gcd(u, p-1) = \prod_{i=1}^{K} p_i^{c_i},$$

then this rewrites as

$$S(t) = \{u \in \mathbb{Z}_{p-1} : \gcd(u, p-1) = \prod_{i=1}^{K} p_i^{c_i}, 0 \leq c_i \leq e_i - \tilde{e}_i\},$$

which, substituted into (8.1), yields

$$R(t) = \sum_{c_1=0}^{e_1 - \tilde{e}_1} \cdots \sum_{c_K=0}^{e_K - \tilde{e}_K} \sum_{b \in T(c_1, \dots c_K)} \prod_{i=1}^{K} p_i^{c_i}$$

$$= \sum_{c_1=0}^{e_1 - \tilde{e}_1} \cdots \sum_{c_K=0}^{e_K - \tilde{e}_K} \prod_{i=1}^{K} p_i^{c_i} \sum_{b \in T(c_1, \dots c_K)} 1$$

$$= \sum_{c_1=0}^{e_1 - \tilde{e}_1} \cdots \sum_{c_K=0}^{e_K - \tilde{e}_K} \prod_{i=1}^{K} p_i^{c_i} |T(c_1, \dots, c_K)|, \tag{8.5}$$

where

$$T(c_1, \dots, c_K) := \{u \in \mathbb{Z}_{p-1} : \gcd(u, p-1) = \prod_{i=1}^{K} p_i^{c_i}\}. \tag{8.6}$$

Lemma 8.3. *For $b \in \mathbb{Z}$ it holds that $b \in T(c_1, \ldots, c_K)$ iff $b \in \mathbb{Z}_{p-1}$ and $m(\alpha^b) = \prod_{i=1}^{K} p_i^{e_i - c_i}$.*

Proof: If we substitute

$$\prod_{i=1}^{K} p_i^{c_i} = \frac{p-1}{\prod_{i=1}^{K} p_i^{e_i - c_i}}$$

into (8.6), we get

$$T(c_1, \ldots, c_K) = \{u \in \mathbb{Z}_{p-1} : \gcd(u, p-1) = \frac{p-1}{\prod_{i=1}^{K} p_i^{e_i - c_i}}\}.$$

This implies that $b \in T(c_1, \ldots, c_K)$ iff $b \in \mathbb{Z}_{p-1}$ and

$$\gcd(b, p-1) = \frac{p-1}{\prod_{i=1}^{K} p_i^{e_i - c_i}}. \tag{8.7}$$

From relation (8.3), (8.7) holds iff

$$m(\alpha^b) = \prod_{i=1}^{K} p_i^{e_i - c_i}. \tag{8.8}$$

So $b \in T(c_1, \ldots, c_K)$ iff $b \in \mathbb{Z}_{p-1}$ and (8.8) holds. $\square$

Proof of Theorem 8.1: Lemma 8.3 yields that $|T(c_1, \ldots, c_K)|$ is exactly the number of elements in $\mathbb{Z}_p^*$ with multiplicative order $\prod_{i=1}^{K} p_i^{e_i - c_i}$. It is known (see e.g. Lidl, Niederreiter (1986), Theorem 1.15) that this number is $\varphi(\prod_{i=1}^{K} p_i^{e_i - c_i})$ (φ denoting the Euler totient function). If we substitute this into (8.5) and use the multiplicativity of the Euler totient function φ for relatively prime elements and the fact that $\varphi(p^e) = p^{e-1}(p-1)$, we obtain

$$R(t) = \sum_{c_1=0}^{e_1-\tilde{e}_1} \cdots \sum_{c_K=0}^{e_K-\tilde{e}_K} \prod_{i=1}^{K} p_i^{c_i} \prod_{i=1}^{K} \varphi(p_i^{e_i-c_i})$$

$$= \prod_{i=1}^{K} \sum_{c_i=0}^{e_i-\tilde{e}_i} p_i^{c_i} \varphi(p_i^{e_i-c_i})$$

$$= \prod_{i=1}^{K} (p_i^{e_i} \delta(\tilde{e}_i) + \sum_{c_i=0}^{e_i-\tilde{e}_i-\delta(\tilde{e}_i)} p_i^{c_i} \varphi(p_i^{e_i-c_i}))$$

$$= \prod_{i=1}^{K} (p_i^{e_i} \delta(\tilde{e}_i) + \sum_{c_i=0}^{e_i-\tilde{e}_i-\delta(\tilde{e}_i)} p_i^{c_i} p_i^{e_i-c_i-1}(p_i-1))$$

$$= \prod_{i=1}^{K} (p_i^{e_i} \delta(\tilde{e}_i) + p_i^{e_i-1}(p_i-1) \sum_{c_i=0}^{e_i-\tilde{e}_i-\delta(\tilde{e}_i)} 1)$$

$$= \prod_{i=1}^{K}(p_i^{e_i}\delta(\tilde{e}_i) + p_i^{e_i-1}(p_i - 1)(e_i - \tilde{e}_i - \delta(\tilde{e}_i) + 1))$$

$$= \prod_{i=1}^{K} p_i^{e_i-1}((p_i - 1)(e_i - \tilde{e}_i + 1) + \delta(\tilde{e}_i)).\square$$

By a simple calculation one deduces from Theorem 8.1:

Corollary 8.1. *If Alice and Bob choose their private keys* X_A, *resp.* X_B, *independently and uniformly at random in* $\mathbb{Z}_{p-1}$, *then*

$$P(\alpha^{X_A X_B} = \alpha^t) = \frac{1}{p-1} \prod_{i=1}^{K}(\frac{p_i - 1}{p_i}(e_i - \tilde{e}_i + 1) + \frac{\delta(\tilde{e}_i)}{p_i}). \qquad (8.9)$$

Let p_{min} resp. p_{max} be the minimum, resp. maximum, possible value (over all $t \in \mathbb{Z}_{p-1}$) of the expression $P(\alpha^{x_A x_B} = \alpha^t)$. From Corollary 8.1 one sees that the minimum value is attained if $e_i = \tilde{e}_i$ for all $i = 1, 2, \ldots, K$, i.e. if the $m(\alpha^t) = p - 1$. This yields

Corollary 8.2.

$$p_{min} = \frac{1}{p-1} \prod_{i=1}^{K} \frac{p_i - 1}{p_i} \approx \frac{1}{p-1}.$$

This means that p_{min} is smaller than the average key probability by only a small factor. On the other hand, the probability $P(\alpha^{x_A x_B} = \alpha^t)$ becomes maximum if $\tilde{e}_1 = \tilde{e}_2 = \ldots \tilde{e}_K = 0$, i.e., if $m(\alpha^t) = 1$. Hence

Corollary 8.3.

$$p_{max} = \frac{1}{p-1} \prod_{i=1}^{K}(e_i \frac{p_i - 1}{p_i} + 1).$$

One can also show:

Corollary 8.4. *If Alice and Bob choose their private keys independently and uniformly in* $\mathbb{Z}_{p-1}^*$, *then the Diffie-Hellman keys are uniformly distributed in* $\mathbb{Z}_{p-1}^*$, *i.e.,*

$$P(\alpha^{X_A X_B} = \alpha^t) = \frac{1}{\varphi(p-1)} \qquad (t \in \mathbb{Z}_{p-1}^*)$$

and zero else.

Proof: Let

$$R^*(t) := \{(x_A, x_B) \in (\mathbb{Z}_{p-1}^*)^2 : \alpha^{x_A x_B} = \alpha^t\}$$
$$= \{(x_A, x_B) \in (\mathbb{Z}_{p-1}^*)^2 : x_A x_B = t(mod.(p - 1))\}$$
$$= \{(x_A x_a^{-1}t) : x_A \in \mathbb{Z}_{p-1}^*\}.$$

So

$$|R^*(t)| = \begin{cases} 0 & : \quad t \notin \mathbb{Z}_{p-1}^* \\ \varphi(p-1) & : \quad t \in \mathbb{Z}_{p-1}^*. \end{cases} \tag{8.10}$$

Furthermore,

$$P(\alpha^{X_A X_B} = \alpha^t) = (\frac{1}{\varphi(p-1)})^2 |R^*(t)|. \tag{8.11}$$

Substituting (8.10) into (8.11) yields the assertion. $\square$

8.3 Strong Primes

In accordance with Massey, Waldvogel (1993), paragraph 4, we will call a prime p a strong prime if it is of the form $p = 2q + 1$ with q prime. From Corollaries 8.2 and 8.3 we obtain, when p is a large strong prime:

$$p_{min} = \frac{1}{p-1} \cdot \frac{1}{2} \cdot \frac{q-1}{q} \approx \frac{1}{p-1} \cdot \frac{1}{2}$$

and

$$p_{max} = \frac{1}{p-1} \cdot \frac{3}{2} \cdot (\frac{q-1}{q} + 1) \approx \frac{1}{p-1} \cdot 3.$$

So in this case, p_{min} and p_{max} are of the same order of magnitude (namely the average key probability). One can also show the following relationship with the entropy:

Corollary 8.5. *If p is a strong prime, then*

$$\log_2(p-1) - 2 < H(\alpha^{X_A X_B}) \leq \log_2(p-1). \tag{8.12}$$

Proof: With the aid of Corollary 8.3 we calculate

$$H(\alpha^{X_A X_B}) \geq - \sum_{t \in \mathbb{Z}_{p-1}} P(\alpha^{X_A X_B} = \alpha^t) \log_2 p_{max}$$

$$= - \log_2 p_{max} \sum_{t \in \mathbb{Z}_{p-1}} P(\alpha^{X_A X_B} = \alpha^t)$$

$$= - \log_2 p_{max}$$

$$= - \log_2(\frac{1}{p-1} \prod_{i=1}^{K} (e_i \frac{p_i - 1}{p_i} + 1))$$

$$= \log_2(p-1) - \sum_{i=1}^{K} \log_2(e_i \frac{p_i - 1}{p_i} + 1)$$

$$> \log_2(p-1) - \sum_{i=1}^{K} \log_2(e_i + 1).$$

This proves the left member of inequality (8.12). The right member is trivial, since it indicates the maximum possible entropy.□

This shows that for large strong primes, the entropy of the Diffie-Hellman key is practically maximum possible, or - in other words - to use strong primes is very good. Without proof we state Corollary 4 of Massey, Waldvogel (1993), which indicates which primes p are worst in the sense that they give large values of p_{max}:

Theorem 8.2. *If $p - 1$ has the prime factorization $p - 1 = \prod_{i=1}^{K} p_i^{e_i}$, then an approximate upper bound for p_{max} is given by the expression*

$$\frac{1}{p-1} \left(\frac{\log(p-1)}{\kappa}\right)^{\kappa} \prod_{i=1}^{[\kappa]} \frac{1}{\log q_i},$$

where q_i denotes the i-th prime,

$$\kappa :\approx \frac{\log(p-1)}{e(\log\log(p-1) - 1)} - 1,$$

$$K :\approx \kappa,$$

$$p_i = q_i,$$

$$e_i \approx \frac{\log_{q_i}(p-1)}{\kappa},$$

and $[x]$ means the rounded value of the real number x to the next integer.

The proof of Theorem 8.2 makes use of the Prime number Theorem, which states that

$$k \sim \frac{q_k}{\log q_k} \qquad (k \to \infty).$$

(In fact, Čebyšev's weak form of it suffices.)

For $p \to \infty$, it was shown in Canetti et al. (1999) that the Diffie-Hellman trjples $(\alpha^{X_A}, \alpha^{X_B}, \alpha^{X_A X_B})$ (α a primitive root modulo p) are uniformly distributed in the sense of Weyl, i.e. interpreted (in the standard way) as elements of the 3-dimensional unit cube $[0, 1]^3$. Their proof is based on estimates for exponential sums and the number of solutions of exponential equations.

9 Differential Cryptanalysis

9.1 The Principle

So-called differential cryptanalysis belongs to the class of chosen-plaintext attacks and was invented by Biham and Shamir (1991). It is a method of cryptanalysis for block ciphers (in contrast to stream ciphers). (In order to avoid misunderstandings from the beginning, note that the term "differential" is used because differences of elements of a commutative group G will be compared and it has nothing to do with calculus!) Let us describe the setting in detail. An r-round block cipher is an encrpytion algorithm that works as follows: For the first round, given an input $X(1)$ and a round key $Z^{(1)}$, the (deterministic) "enciphering function" f produces an output $Y(1) = f(X(1), Z^{(1)})$. The output of the first round is used as input for the second round $X(2) := Y(1)$, and as output of the second round we get $Y(2) = f(X(2), Z^{(2)})$, etc. The final output of the algorithm will be the output of the r-th round $Y(r)$. Here, all occurring data are blocks of a certain length whose elements belong to some finite abelian group, in practice often some residue ring. The model assumption will be that all round keys $Z^{(1)}, Z^{(2)}, \ldots Z^{(r)}$ are chosen as independent uniformly distributed random variables, for general, only in this case do reasonable theoretic results become available. But interestingly enough, in practice, it seems to work as well or even better when the round keys are determined by some key schedule for a "small" overall key. Now the idea of differential cryptanalysis is that if one takes pairs of round inputs $(X(i), X^*(i))$ and compares them with the round output pairs $(Y(i), Y^*(i))$, often there are relations between their differences $\Delta X(i) := X(i) - X^*(i)$ and $\Delta Y(i) := Y(i) - Y^*(i)$ that allow as to infer information on the round key $Z^{(i)}$. Informally speaking, the enciphering function f is called cryptographically weak if for given $\Delta Y(r-1)$, $Y(r)$, $Y^*(r)$ for a relatively small number of input pairs $(X(1), X^*(1))$, one can "easily" find the round key $Z^{(r)}$ or at least some information about it. A pair of differences (α, β) considered as values of a pair of first-round input and i-th-round output $(\alpha, \beta) = (\Delta X(1), \Delta Y(i))$ is termed an i-round differential (or characteristic). Differential cryptanalysis is successful if there are differentials that are significantly more probable than others if the round keys $Z^{(1)}, Z^{(2)}, \ldots, Z^{(r-1)}$ are chosen uniformly at random. Now the differential attack proceeds as follows:

D. Neuenschwander: Prob. and Stat. Methods in Cryptology, LNCS 3028, pp. 115-123, 2004.
© Springer-Verlag Berlin Heidelberg 2004

- Choose an $(r-1)$-round differential (α, β) for which the conditional probability $P(\Delta Y(r-1) = \beta \mid \Delta X(1) = \alpha)$ is relatively large.
- Take a plaintext $X(1)$ chosen uniformly at random and encrypt $X(1)$ and $X^*(1) := X(1) + \alpha$ to get the ciphertexts $Y(r)$ and $Y^*(r)$.
- Assume that β is the true difference $\Delta Y(r-1)$. Find all values of the round key $Z^{(r)}$ that are consistent with r-round input difference β and output difference $\Delta Y(r) = Y(r) - Y^*(r)$.
- Repeat the two preceding steps until some possible $Z^{(r)}$ appears significantly more frequently than all the others. Then use this value as a guess for the r-th round key.
- Do all these steps iteratively for $r-1, r-2, \ldots, 1$.

The creative act needed to mount a differential attack lies in the first step, i.e., to find a significantly more probable differential. This is why information about the distribution of differentials is important. We will treat this question in the next section.

Fortunately, by the following theorem due to Lai, Massey, and Murphy, there is a lower bound on the complexity of a differential attack. Here, "complexity" means the number of times an encryption of a chosen plaintext pair must be made.

Theorem 9.1. *Let G be an abelian group (in particular, a residue ring), N be the block length, and put*

$$p_{\max} := \max_{\alpha, \beta \in G} \{ P(\Delta Y(r-1) = \beta \mid \Delta X = \alpha \}.$$

Then the average complexity C of the differential cryptanalysis has the following lower bound:

$$C \geq \frac{2}{p_{\max} - \frac{1}{2^N - 1}}.$$

Proof: If the attack succeeds, then the anticipated value β has to occur at least once more than a uniformly randomly chosen other β'. In K pairs of encryptions, on the average β occurs $K p_{\max}$ and β' occurs $K(2^N - 1)^{-1}$ times. Thus

$$K p_{\max} - K \frac{1}{2^N - 1} \geq 1,$$

which, by resolving with respect to K, yields the assertion. $\square$

So, the smaller $p_{\max}$ (i.e., the less there are significantly more probable differentials), the bigger the complexity becomes.

Of course, the cardinal question here is how to design a cipher that, against differential cryptanalysis, is reasonably secure. It turns out that for this, the notion of a Markov cipher seems to be a natural condition. The following definition is due to Lai, Massey, and Murphy.

Definition 9.1. *An r-round iterated block cipher is called a Markov cipher if, when the first round key $Z^{(1)}$ is chosen uniformly at random, then the probability*

$$P(\Delta Y(1) = \beta \quad | \quad \Delta X(1) = \alpha, X(1) = \gamma)$$

is independent of γ for all α, β, γ.

To be exact, we need the model assumption of stochastic equivalence:

Definition 9.2. *The assumption of stochastic equivalence means that $P(\Delta Y(r - 1) = \beta \mid \Delta X(1) = \alpha)$ has the same value for fixed round keys $Z^{(1)}, Z^{(2)}, \ldots, Z^{(r-1)}$ as if these round keys $Z^{(i)}$ $(i = 1, 2, \ldots, r - 1)$ were independent and uniformly distributed.*

As Biham and Shamir have shown, e.g., DES is a Markov cipher. The relation of the above definition to Markov chains is the following theorem due to Lai, Massey, and Murphy:

Theorem 9.2. *If in an r-round Markov cipher, all round keys are chosen independently and uniformly at random, then $\{\Delta Y(i)\}_{0 \le i \le r}$ is a Markov chain.*

(Here, the term "Markov chain" will always mean "homogeneous" Markov chain.)

Proof of Theorem 9.2: We have

$$P(\Delta Y(1) = \beta_1, \Delta Y(2) = \beta_2, \ldots, \Delta Y(r) = \beta_r \quad | \quad \Delta Y(0) = \beta_0)$$
$$= \prod_{i=1}^{r} P(\Delta Y(i) = \beta_i \quad | \quad \Delta Y(0) = \beta_0, \Delta Y(1) = \beta_1, \ldots, \Delta Y(i - 1) = \beta_{i-1}).$$

However,

$$P(\Delta Y(i) = \beta_i \quad | \quad \Delta Y(0) = \beta_0, \Delta Y(1) = \beta_1, \ldots, \Delta Y(i - 1) = \beta_{i-1})$$
$$= \sum_{\gamma \in G} P(\Delta Y(i) = \beta_i, Y(i - 1) = \gamma \quad | \quad \Delta Y(0) = \beta_0, \Delta Y(1) = \beta_1, \ldots,$$
$$\Delta Y(i - 1) = \beta_{i-1})$$

and

$$P(\Delta Y(i) = \beta_i, Y(i - 1) = \gamma \quad | \quad \Delta Y(0) = \beta_0, \Delta Y(1) = \beta_1, \ldots, \Delta Y(i - 1) = \beta_{i-1})$$
$$= P(Y(i - 1) = \gamma \quad | \quad \Delta Y(0) = \beta_0, \Delta Y(1) = \beta_1, \ldots, \Delta Y(i - 1) = \beta_{i-1})$$
$$\cdot P(\Delta Y(i) = \beta_i \quad | \quad Y(i - 1) = \gamma, \Delta Y(0) = \beta_0, \Delta Y(1) = \beta_1, \ldots, \Delta Y(i - 1) = \beta_{i-1}).$$

By the independence of the round keys and the definition of a Markov cipher, we have

$$P(\Delta Y(i) = \beta_i \quad | \quad Y(i - 1) = \gamma, \Delta Y(0) = \beta_0, \Delta Y(1) = \beta_1, \ldots,$$
$$\Delta Y(i - 1) = \beta_{i-1})$$
$$= P(\Delta Y(i) = \beta_i \quad | \quad Y(i - 1) = \gamma, \Delta Y(i - 1) = \beta_{i-1})$$
$$= P(\Delta Y(i) = \beta_i \quad | \quad \Delta Y(i - 1) = \beta_{i-1}).$$

So we get

$$P(\Delta Y(i) = \beta_i \mid \Delta Y(0) = \beta_0, \Delta Y(1) = \beta_1, \ldots, \Delta Y(i-1) = \beta_{i-1})$$
$$= P(\Delta Y(i) = \beta_i \mid \Delta Y(i-1) = \beta_{i-1})$$
$$\cdot \sum_{\gamma \in G} P(Y(i-1) = \gamma \mid \Delta Y(0) = \beta_0, \Delta Y(1) = \beta_1, \ldots, \Delta Y(i-1) = \beta_{i-1}).$$

Since the latter sum adds up to 1, we finally obtain

$$P(\Delta Y(1) = \beta_1, \Delta Y(2) = \beta_2, \ldots, \Delta Y(r) = \beta_r \mid \Delta Y(0) = \beta_0)$$
$$= \prod_{i=1}^{r} P(\Delta Y(i) = \beta_i \mid \Delta Y(i-1) = \beta_{i-1}).$$

Homogeneity follows from the fact that all round keys have the same (uniform) distribution. $\square$

Lemma 9.1. *For any Markov cipher, the uniform distribution on $G^N \backslash \{e\}$ is a stationary distribution of the Markov chain $\{\Delta Y(i)\}_{0 \le i \le r}$.*

Proof: Put $Y(i) = X(i+1) = e$ and choose $Y^*(i) = X^*(i+1)$ uniformly on $G^N \backslash \{e\}$ at random. Then, since the cipher is Markov, the random variable $\Delta Y(i)$ obeys itself a uniform distribution on $G^N \backslash \{e\}$. For any fixed $(i+1)$-th round key $z = Z^{(i+1)}$, the random variable $Y^*(i+1) = f(X^*(i+1), z)$ is uniformly distributed on $G^N \backslash \{f(e,z)\}$, since $f(.,z)$ is invertible. Thus for fixed z, the random variable $\Delta Y(i+1)$ is uniformly distributed over $G^N \backslash \{e\}$. Hence the same is also true without conditioning on $Z^{(i+1)}$. $\square$

A stronger notion than a stationary probability measure of a Markov chain is the concept of a so-called steady-state distribution. This means the following:

Definition 9.3. *The Markov chain $\{\Delta Y(i)\}_{i \ge 0}$ is said to have the steady-state distribution π if for all $\alpha, \beta, \Delta Y(.)$ it holds that*

$$P(\Delta Y(i) = \beta \mid \Delta Y(0) = \alpha) \to \pi(\beta) \qquad (i \to \infty).$$

If a Markov chain has a steady-state distribution, then this is its unique stationary distribution. Now by the following theorem due to Lai, Massey, and Murphy, it turns out that Markov ciphers having a steady-state distribution are "immune" to differential cryptanalysis.

Theorem 9.3. *Under the assumption of stochastic equivalence, Markov ciphers having a steady-state distribution are (asymptotically as the number of rounds tends to infinity) immune to differential cryptanalysis (in the sense that the average complexity tends to ∞).*

Proof: From Theorem 9.1 and the fact that from Lemma 9.1 $p_{\max} \to \frac{1}{2^N - 1}$, we have $C \to \infty$ as the number of rounds tends to infinity. $\square$

9.2 The Distribution of Characteristics

As mentioned in Section 9.1, it is important to know the distribution of differentials under the null hypothesis that the keys are chosen uniformly at random. In this section, we will work in somewhat more generality in the sense that we will look at additive characteristics in powers of groups $\mathbb{Z}_q$. For $q = 2$ this is just classical differential cryptanalysis, since in this case, $+$ and $-$ are the same. Let $q \in \mathbb{N}$, $q \geq 2$ and fix $\Delta X, \Delta Y \in (\mathbb{Z}_q)^m$. Let π be a uniformly distributed random permutation of $(\mathbb{Z}_q)^m$ (which occurs due to a randomly chosen key K) and consider the random variable $\Lambda_\pi(\Delta X, \Delta Y)$ giving the number of (unordered) pairs $\{X, X'\} \subset (\mathbb{Z}_q)^m$ of plaintexts X, X' such that $X + X' = \Delta X$ and $Y + Y' = \Delta Y$, where $Y = \pi(X)$, $Y' = \pi(X')$ are the corresponding ciphertexts.

We begin with an elementary algebraic lemma, whose proof follows from standard properties of linear diophantine equations.

Lemma 9.2. *Let $q \in \mathbb{N}$, $q \geq 2$ and $k \in \mathbb{Z}$. Consider the equation*

$$2x = k \quad (mod.\ q) \quad (x \in \mathbb{Z}_q). \tag{9.1}$$

If q is odd, then (9.1) has exactly one solution mod. q. If q is even and k is odd, then (9.1) has no solution. If q and k are both even, then (9.1) has exactly two solutions mod. q.

The next lemma is the so-called "pairing theorem", a combinatorial assertion.

Lemma 9.3. *Let $A = \{a_1, a_2, \ldots, a_{2d}\}$ and $B = \{b_1, b_2, \ldots, b_{2d}\}$ be alphabets with $2d$ distinct elements. Assume Π_A and Π_B are sets of unordered pairs such that a_i (resp. b_i) occurs in exactly one pair of Π_A resp. Π_B ($i = 1, 2, \ldots, 2d$). Denote by $\Psi(d)$ the number of bijections $\psi : A \to B$ such that for pairs $\{a_i, a_j\} \in \Pi_A$ we have $\{\psi(a_i), \psi(a_j)\} \notin \Pi_B$. Then we have*

$$\Psi(d) = \sum_{k=0}^{d} (-1)^k \binom{d}{k}^2 2^k k! (2d - 2k)!. \tag{9.2}$$

Proof: We order the set Π_B as $\{\{b'_i, b'_{i+d}\}\}_{1 \leq i \leq d}$ and let $P(i)$ be the number of bijections $\psi : A \to B$ that map some pair of Π_A to the pair $\{b'_i, b'_{d+i}\}$ in Π_B, i.e.,

$$P(i) := \{\psi : \psi(a) = b'_i, \psi(a') = b'_{d+i}, \{a, a'\} \in \Pi_A\}.$$

By the principle of inclusion-exclusion we get

$$\Psi(d) = (2d)! - |\bigcup_{1 \leq j \leq d} P(j)|$$

$$= (2d)! + \sum_{\emptyset \neq S \subset \{1,2,\ldots,d\}} (-1)^{|S|} |\bigcap_{j \in S} P(j)|. \tag{9.3}$$

If we define (by symmetry)

$$
\begin{aligned}
P(d,k) &:= P(1,2,\ldots,k) \\
&= P(i'_1, i'_2, \ldots, i'_k) \\
&:= |\bigcap_{1\leq j\leq k} P(i'_j)|,
\end{aligned}
\tag{9.4}
$$

we obtain the relation

$$
\Psi(d) = (2d)! + \sum_{k=1}^{d} (-1)^k \binom{d}{k} P(d,k).
\tag{9.5}
$$

Now we order Π_A in the same way as Π_B, i.e., as $\{\{a'_i, a'_{i+d}\}\}_{1\leq i\leq d}$. Then $P(d,k)$ can be interpreted as the number of functions $\psi : A \to B$ for which there are exactly k pairs $\{a''_i, a''_{i+d}\}$ from Π_A such that $\psi(a''_i) = b'_i$ and $\psi(a''_{i+d}) = b''_{i+d}$ ($i = 1, 2, \ldots, k$). By elementary combinatorial considerations, it turns out that there exist $\binom{d}{k}$ ways to select the k pairs $\{a''_i . a''_{i+d}\}$ from Π_A, then $k!$ possibilities to assign the pairs $\{a''_i, a''_{i+d}\}$ to the pairs $\{b'_i, b'_{i+d}\}$, and at the end 2^k ways to assign $\{a'_i, a'_{i+d}\}$ to a particular pair in Π_B. Finally, the number of ways to assign the elements of

$$
A\backslash \bigcup_{1\leq i\leq k} \{a''_i, a''_{i+d}\}
$$

is given by $(2d - 2k)!$. So

$$
P(d,k) = \binom{d}{k} 2^k k! (2d - 2k)!
\tag{9.6}
$$

and the assertion follows from (9.5). $\square$

Theorem 9.4. *Suppose $q \in \mathbb{N}$, $q \geq 2$, q even. Let $\Delta X = (\triangle X_1, \triangle X_2, \ldots, \triangle X_m)$, $\Delta Y = (\triangle Y_1, \triangle Y_2, \ldots, \triangle Y_m) \in (\mathbb{Z}_q)^m$ such that at least one of the $\triangle X_i$ and at least one of the $\triangle Y_i$ is odd. Then the distribution of the random variable $\Lambda_\pi(\Delta X, \Delta Y)$ tends to the Poisson distribution given by*

$$
P(H = k) = e^{-1/2} 2^{-k} k!^{-1} \quad (k \in \mathbb{N}^0) \quad (m \to \infty).
\tag{9.7}
$$

Proof: If $X + X' = \Delta X$, then from Lemma 9.2 it is not possible that $X = X'$. Thus, from Lemma 9.3 the number of permutations π of $(\mathbb{Z}_q)^m$ such that if $X + X' = \Delta X$ then $\pi(X) + \pi(X') \neq \Delta Y$ is given by $\Psi(q^m)$, where

$$
\Psi(2d) := \sum_{k=0}^{d} (-1)^k \binom{d}{k}^2 2^k k! (2d - 2k)!.
\tag{9.8}
$$

We get (for u in a neighborhood of 1), for the generating function of $\Delta_\pi(\Delta X, \Delta Y)$,

$$E(u^{\Lambda_\pi(\Delta X, \Delta Y)}) = \sum_{k=0}^{q^m/2} \binom{q^m/2}{k}^2 \frac{u^k k! 2^k \Psi(q^m - 2k)}{q^m!}. \tag{9.9}$$

[From the definition of $P(d, k)$ and (9.6) it follows that the expression

$$\frac{P(d, k)\Psi(d - k)}{(2d - 2k)!}$$

denotes the number of functions ψ that take exactly k pairs from Π_A to (b'_i, b'_{i+k}) (with some abuse of notation). Then the number of permutations of $(\mathbb{Z}_q)^m$ for which k pairs of sum ΔX can be mapped into k fixed pairs of difference ΔY is given by the expression

$$\sum_{S \subset (\mathbb{Z}_q)^{m-1}, |S|=k} \frac{P(q^{m-1}, k)\Psi(q^{m-1} - k)}{(q^m - 2k)!} = \binom{q^{m-1}}{k}.$$

From (9.6), we get

$$|\{\pi : \Lambda(\Delta X, \Delta Y) = k\}| = \binom{q^{m-1}}{k} \frac{P(q^{m-1}, k)\Psi(q^{m-1} - k)}{(q^m - 2k)!}$$

$$= \binom{q^{m-1}}{k}^2 k! 2^k \Psi(q^{m-1} - k).$$

The assertion follows.]
We have to determine the limit of expression (9.9) as $m \to \infty$. For this, we first calculate the limit of (9.8) (with $d = q^m/2$) as $m \to \infty$. Put

$$T(m, k) := (-1)^k \binom{q^m/2}{k}^2 2^k k! (q^m - 2k)!. \tag{9.10}$$

Then for the ratio of two consecutive (with respect to k) such expressions

$$\begin{aligned}
\frac{T(m.k + 1)}{T(m, k)} &= -2 \frac{(q^m/2 - k)^2}{(k + 1)(q^m - 2k)(q^m - 2k - 1)} \\
&= -2 \frac{(q^m/2 - k)^2}{4(k + 1)(q^m/2 - k)^2 - (k + 1)(q^m - 2k)} \\
&= -(2(k + 1)(1 - \frac{1}{q^m - 2k}))^{-1}. \tag{9.11}
\end{aligned}$$

So asymptotically by successive multiplications of the terms (9.11) we obtain

$$\frac{T(m, k)}{T(m, 0)} = \frac{(-1)^k}{2^k k!} \varepsilon_k, \tag{9.12}$$

where $\varepsilon_k \to 1$ when $k \in o(q^m/2)$ $(m \to \infty)$, $\varepsilon_0 := 1$, and $\varepsilon_k = O(q^m)$ when $k \notin o(q^m/2)$ $(m \to \infty)$; hence (for Ψ as in (9.8)) $T(m,0)$ behaves asymptotically as $\Psi(q^m)$ in the sense that

$$\frac{\Psi(q^m)}{T(m,0)} \to \frac{1}{\sqrt{e}} \qquad (m \to \infty). \tag{9.13}$$

Now, in view of calculating the generating function with variable u, we define an analogous expression as (9.10) (for u in a neighborhood of 1) but including an additional term u^k, replacing the last factorial in (9.10) by Ψ, and without change of sign $(-1)^k$:

$$T_u(m,k) := \binom{q^m/2}{k}^2 u^k k! 2^k \Psi(q^m - 2k). \tag{9.14}$$

Then, we get the ratio

$$\frac{T_u(m,k)}{T_u(m,1)} = \frac{1}{k!} \left(\frac{u}{2}\right)^{-(k-1)}, \tag{9.15}$$

hence the generating function has asymptotic behavior

$$E(u^{\Lambda_\pi(\Delta X, \Delta Y)}) \sim \frac{2e^{u/2} T_u(m,1)}{u q^m!} \qquad (m \to \infty). \tag{9.16}$$

From (9.14) and an elementary estimation, the term $T_u(m,1)$ in (9.16) behaves as

$$T_u(m,1) \sim 2(q^m/2)^2 u \Psi(q^m - 2)$$
$$\sim q^{2m} 2^{-1} \frac{u}{\sqrt{e}} (q^m - 2)! \qquad (m \to \infty), \tag{9.17}$$

thus from (9.16) and (9.17)

$$E(u^{\Lambda_\pi(\Delta X, \Delta Y)}) \sim \frac{e^{u/2} q^{2m} (q^m - 2)!}{\sqrt{e} q^m!}$$
$$\to e^{(1/2)(u-1)} \qquad (m \to \infty). \square \tag{9.18}$$

Next, let us treat the case where q is odd. In this situation, for any ΔX there is, from Lemma 9.2, exactly one X such that $2X = \Delta X$.

Theorem 9.5. *Suppose $q \in \mathbb{N}$, $q \geq 3$, q odd. Let $\Delta X, \Delta Y \in (\mathbb{Z}_q)^m$. Then the distribution of the random variable $\Lambda_\pi(\Delta X, \Delta Y)$ tends weakly to the distribution (9.7) as $m \to \infty$.*

Proof: Let $X \in \mathbb{Z}_q$ be the unique solution of (9.1) and let $X_0 = (X, X, \ldots, X)(\in (\mathbb{Z}_q)^m)$. In contrast to the proof of Theorem 9.4, one has to count those cases where $\pi(X_0) = X_0$ and $\pi(X_0) \neq X_0$ separately, which yields as analogue of (9.9) the expression

$$E(u^{\Lambda_\pi(\Delta X, \Delta Y)})$$

$$= \frac{1}{q^m!} \left(\sum_{k=0}^{(q^m-1)/2} \binom{(q^m-1)/2}{k} u^k k! 2^k \Psi(q^m - 1 - 2k) \right.$$

$$\left. + (q^m - 1) \sum_{k=1}^{(q^m-1)/2} \binom{(q^m-1)/2}{k-1} u^{k-1}(k-1)! 2^{k-1} \Psi(q^m + 1 - 2k) \right). \quad (9.19)$$

Now the same type of limit procedure as in the proof of Theorem 9.4, applied separately to both sums on the right-hand side of (9.19), yields the result. □
What remains is the case where q and all $\triangle X_i$, $\triangle Y_i$ are even. Here, by Lemma 9.2, equation (9.1) has exactly 2 solutions, hence we get (by analogy to the foregoing cases) the following result:

Theorem 9.6. *Suppose $q \in \mathbb{N}$, $q \geq 2$, q even. Let $\Delta X = (\triangle X_1, \triangle X_2, \ldots, \triangle X_m)$, $\Delta Y = (\triangle Y_1, \triangle Y_2, \ldots, \triangle Y_m) \in (\mathbb{Z}_q)^m$ such that all $\triangle X_i$ and all $\triangle Y_i$ are even. Then the distribution of the random variable $\Lambda_\pi(\Delta X, \Delta Y)/2^m$ tends weakly to the distribution (9.7) as $m \to \infty$.*

10 Semantic Security

At the end of Chapter 1, in connection with the One-Time Pad, we discussed the notion of perfect secrecy. The effect of perfect secrecy is that the
adversary, even if he has unlimited resources, can not gain any information
about the plaintext from the ciphertext, except its length (if this is not a
known parameter). The fact that any cryptosystem leaks the information
about the length of the plaintext will be proved below (Theorem 10.1). Another notion, related to perfect secrecy, is that of so-called semantic security.
Roughly speaking, semantic security is a polynomially bounded variant of
perfect security, i.e., one assumes that the adversary has only polynomially
bounded resources.

Let us fix definitions and notations.

In this chapter, we will use the term "random variable" in a somewhat non-
classical sense:

Definition 10.1. *A random variable is a sequence* $\{X_n\}_{n \geq 1}$ *of random variables* X_n *in the classical sense defined over some common probability space*
$(\Omega, \mathcal{A}, P)$ *such that there is a polynomial* Q *so that (for all* n*)* X_n *ranges
over* $\mathbb{B}^{Q(n)-1}$*. The random variable is called polynomial-time, if there exists
a probabilistic polynomial-time algorithm* A *such that*

$$P(A(1^n) = x) = P(X_n = x)$$

(1^n *means the string* $(1, 1, \ldots, 1) \in \mathbb{B}^n$*).*

Definition 10.2. *A cryptosystem is a triple* (G, E, D) *of probabilistic polynomial-time algorithms such that*

- *For the input* 1^n*, algorithm* G *(the key generator) outputs two bitstrings*
  $G_1(1^n)$ *and* $G_2(1^n)$ *both of length* n*.*
- *For every pair* (e, d) *of encrpytion/deciphering keys in the range of* $G(1^n)$*,
  the encryption algorithm* E *and the deciphering algorithm* D *satisfy, for
  each plaintext* $x \in \mathbb{B}^n$*, the relation*

$$P(D_d(E_e(x)) = x) = 1.$$

- *There exists a polynomial* Q *such that for all* $e, x \in \mathbb{B}^n$*, we have that the
  random variable* $E_e(x)$ *ranges over* $\mathbb{B}^{Q(n)}$*.*

D. Neuenschwander: Prob. and Stat. Methods in Cryptology, LNCS 3028, pp. 125-133, 2004.
© Springer-Verlag Berlin Heidelberg 2004

The last requirement has the disadvantage that it always reveals the length $|x|$ of the plaintext. However, cryptosystems always leak the information about the length of the plaintext, as the following theorem shows:

Theorem 10.1. *Let (G, E, D) be a cryptosystem not necessarily satisfying the length conditions imposed in Definition 10.2. (In particular, E is defined for every possible key and every plaintext and the only restriction on the distribution of the length of the ciphertext produced by E is that it must be polynomial in the length of the inputs to E.) Then this scheme can not hide the length of the plaintext.*

Proof: Let (in accordance with Definition 10.2), e be an encryption key in the range of $G(1^n)$ and consider the random variables $E_e(1^m)$ and $E_e(1^{m+1})$ for some m that is polynomial in $|e|$ (the length of e). If the encryption hides the length of the plaintext, then these two random variables have to be polynomially indistinguishable. Let Q be the polynomial bounding the running time of the encryption algorithm E and let m take the values $|e|, |e| + 1, \ldots, Q(2|e|) + 1$. Then we find that the random variables $E_e(1^{|e|})$ and $E_e(1^{Q(2|e|)+2})$ are polynomially indistinguishable, hence (since $P(|E_e(1^{|e|})| \le Q(2|e|)) = 1$) we have the lower bound

$$P(|E_e(1^{Q(2|e|)+2})| \le Q(2|e|)) > 2/3.$$

But from the fact that the code must be uniquely decipherable, it follows that for at most half of the bitstrings x of length $|x| = Q(2|e|) + 2$ it holds that

$$P(|E_e(x)| \le Q(2|e|) + 2) > 1/3.$$

So there exists an $x \in \mathbb{B}^{Q(2|e|)+2}$ such that $E_e(x)$ and $E_e(1^{Q(2|e|)+2})$ are distinguishable by just measuring their lengths, i.e. in polynomial time. This is a contradiction.□

Now we come to the definition of semantic security.

For a set Σ, the symbol Σ^* denotes the set of all finite sequences of elements of Σ.

Definition 10.3. *A (secret-key) cryptosystem (G, E, D) as in Definition 10.2 is called semantically secure if, for every probabilistic polynomial-time algorithm A, there exists a probabilistic polynomial-time algorithm A' such that for every polynomial-time random variable $\{X_n\}_{n\ge 1}$, every polynomial-time computable function $h : \mathbb{B}^* \to \mathbb{B}^*$, every function $f : \mathbb{B}^* \to \mathbb{B}^*$, every positive constant c, and all sufficiently large n,*

$$P(A(E_{G_1(1^n)}(X_n), h(X_n), 1^n) = f(X_n))$$
$$< P(A'(|X_n|, h(X_n), 1^n) = f(X_n)) + n^{-c}.$$

This definition can be explained as follows: The role of the function h is to provide partial information on X_n to both algorithms, which then try to find

$f(X_n)$. As we shall see in the sequel, the meaning of semantic security is, roughly speaking, that the distribution of the random variables

$$A(E(X_n), h(X_n), 1^n)$$

and

$$A'(|X_n|, h(X_n), 1^n)$$

are close in a certain sense. We will see that if also the function f is supposed to be computable in polynomial time, then the definition of semantic security remains equivalent. In Definition 10.2, we considered secret-key cryptosystems. The corresponding notion of semantic security of a public-key system should be formulated in the sense that the public key is given to the algorithm A as additional input. Evidently, any public-key system that is semantically secure has a probabilistic encryption algorithm (otherwise, look at a random variable X_n that is uniformly distributed on the two-point set $\{0^n, 1^n\}$). Now we will show that semantic security is equivalent to so-called indistinguishable encryption, which means the following:

Definition 10.4. *A (secret-key) cryptosystem (G, E, D) as in Definition 10.2 has hte indistinguishable encryption property if, for every polynomial-time random variable $\{T_n\}_{n \geq 1} = \{(X_n, Y_n, Z_n)\}_{n \geq 1}$ with $|X_n| = |Y_n|$, every probabilistic polynomial-time algorithm A, every positive constant c, and all sufficiently large n, we have that*

$$|P(A(Z_n, E_{G_1(1^n)}(X_n)) = 1) - P(A(Z_n, E_{G_1(1^n)}(Y_n)) = 1)| < n^{-c}.$$

The random variable Z_n has to be interpreted as additional information, on the space of plaintexts, given to algorithm A, which tries to distinguish the encryptions of X_n and Y_n. An analogous remark as we have stated for semantic security of public-key systems holds here: The public key (i.e. $G_1(1^n)$) should be taken as an additional input to the algorithm.

Our goal now will be to show that semantic security and indistinguishable encryption are equivalent properties. Especially the direction that indistinguishable encrpytion implies semantic security is very important, since it often seems easier to prove the former than the latter.

Theorem 10.2. *A (secret-key) cryptosystem as in Definition 10.2 is semantically secure iff it has the property of indistinguishable encryption.*

For the proof of Theorem 10.2 we verify both directions by individual propositions, which are both in fact stronger than the corresponding direction of Theorem 10.2 and thus of a certain own interest.

Proposition 10.1. *Let (G, E, D) have the property of indistinguishable encryption and suppose, furthermore, that $Z_n = X_n Y_n$. Then the system is semantically secure.*

Proof: Let us assume that the system is not semantically secure. We will prove that in this case it has distinguishable encryptions with $Z_n = X_n Y_n$. This will be done by showing that if for some $\{X_n\}_{n\geq 1}$, f, h as in Definition 10.3, and a probabilistic polynomial-time algorithm A, there exists a probabilistic polynomial-time algorithm A' such that if A guesses $f(X_n)$ from the encrypted value $E_{G_1(1^n)}(X_n)$ and $h(X_n)$ better than A' does on input $|X_n|$ and $h(X_n)$, then one can distinguish the encryptions of X_n and $Y_n := 1^{|X_n|}$ (using $Z_n = X_n Y_n$ as auxiliary input). Let A be a probabilistic polynomial-time algorithm that tries to guess partial information (i.e., $f(X_n)$) from the encryption of X_n and the a priori information $h(X_n)$. Namely, on input $E_{G_1(1^n)}(x)$ and $h(x)$, the algorithm A tries to guess $f(x)$. Now we construct a probabilistic polynomial-time algorithm A' that has as good performance but without getting the input $E_{G_1(1^n)}(x)$. This algorithm will run algorithm A with input $E_{G_1(1^n)}(1^{|x|})$ and $h(x)$. We will show that

$$P(A(E_{G_1(1^n)}(X_n), h(X_n), 1^n) = f(X_n))$$
$$< P(A'(|X_n|, h(X_n), 1^n) = f(X_n)) + n^{-c}$$

or, as is equivalent,

$$P(A(E_{G_1(1^n)}(X_n), h(X_n), 1^n) = f(X_n))$$
$$< P(A(E_{G_1(1^n)}(1^{|X_n|}), h(X_n), 1^n) = f(X_n)) + n^{-c}.$$

Assume the contrary and let $c > 0$ be such that the above fails for infinitely many n. Then we have

$$P(X_n \in B_n) \geq \frac{1}{2} n^{-c},$$

where B_n is the set of bitstrings x of length m with the property

$$P(A(E_{G_1(1^n)}(x), h(x), 1^n) = f(x))$$
$$> P(A(E_{G_1(1^n)}(1^n), h(x), 1^n) = f(x)) + \frac{1}{2} n^{-c}.$$

If D_n denotes the set of bitstrings of length m satisfying

$$|P(A(E_{G_1(1^n)}(x), h(x), 1^n) = \xi_x)$$
$$- P(A(E_{G_1(1^n)}(1^n), h(x), 1^n) = \xi_x)| > \frac{1}{2} n^{-c} \tag{10.1}$$

for some ξ_x, then we have

$$P(X_n \in D_n) \geq \frac{1}{2} n^{-c}.$$

We now define a random variable $\{Z_n\}_{n\geq 1} = \{X_n \cdot 1^{|X_n|}\}_{n\geq 1}$ and construct a polynomial-time probabilistic algorithm A_1 that, given auxiliary information

$(X_n, 1^m)$ (where $m := |X_n|$), distinguishes the encryptions of $X_n \in D_n$ and 1^m. The algorithm A_1 is defined as follows:

On input x, 1^m, and $E_e(w)$ (where $w \in \{x, 1^m\}$ and e is in the range of $G_1(1^n)$), the algorithm has two steps. Roughly speaking, the first step consists of checking if $x \in D_n$ and, if yes, then to find a ξ_x satisfying relation (10.1). (This ξ_x can be interpreted as some sort of "witness" for the fact that $x \in D_n$.) Then, in the second step, the algorithm "guesses" the identity of w by checking if $A(E_e(w), h(x), 1^n) = \xi_x$ or not. In detail:

- Ignoring $E_e(w)$, the algorithm first gathers information on the statistics of the random variables $A(E_{G_1(1^n)}(x), h(x), 1^n)$ and $A(E_{G_1(1^n)}(1^m), h(x), 1^n)$ by computing $h(x)$ and running A polynomially often (depending on the desired accuracy and determinable by the following equation (10.3)), each time giving to A as input a randomly computed $E_{G_1(1^n)}(x)$, resp. $E_{G_1(1^n)}(1^m)$, and the data $h(x)$ and 1^n. Put

$$p_{x,v}(\xi) := P(A(E_{G_1(1^n)}(v), h(x), 1^n) = \xi) \qquad (10.2)$$

  and let $\widehat{p}_{x,v}(\xi)$ be a random variable representing the estimator of $p_{x,v}(\xi)$ that is obtained by polynomially many runs, the polynomial again defined such that (10.3) (see infra) holds. If we fix x and v, then with probability at least say $1 - 2^{-n}$, for every possible value ξ we have that $\widehat{p}_{x,v}(\xi)$ is a good estimator for $p_{x,v}(\xi)$ in the sense that

$$|p_{x,v}(\xi) - \widehat{p}_{x,v}(\xi)| < \frac{1}{16} n^{-2c}. \qquad (10.3)$$

If $x \in D_n$, it holds that

$$|p_{x,x}(\xi) - \widehat{p}_{x,1^m}(\xi)| > \frac{1}{2} n^{-c}.$$

So from (10.3) we find that with very high probability, if $x \in D_n$, we can find a ξ^* with

$$|\widehat{p}_{x,x}(\xi^*) - \widehat{p}_{x,1^m}(\xi^*)| > \frac{3}{8} n^{-c}. \qquad (10.4)$$

If such a ξ^* cannot be found (as just mentioned, this occurs with very low probability), then the algorithm A_1 terminates here and outputs (obliviously of $E_e(w)$) the value 1.

- Assume we have found the above-mentioned ξ^*. W.l.o.g. we may assume that

$$\widehat{p}_{x,x}(\xi^*) > \widehat{p}_{x,1^m}(\xi^*) + \frac{3}{8} n^{-c}.$$

  Now algorithm A_1 runs the algorithm $A(E_e(w), h(x), 1^n)$ and gives 1 as output iff A yields output ξ^*.

It remains to analyze the performance of the just-defined algorithm A_1 (i.e., to prove that it is really a probabilistic polynomial-time algorithm) on input $E_e(w)$ (where $w \in \{x, 1^m\}$), 1^m, and x. We have to distinguish 3 possible cases:

- If $x \in D_n$ (this event will be denoted by C_1 in the sequel), w.l.o.g. we may assume that the ξ^* (which has been found with probability at least $1 - 2^{-n}$ say) satisfies

$$p_{x,x}(\xi^*) \geq p_{x,1^m}(\xi^*) + \frac{3}{8}n^{-c}.$$

Then it follows that

$$P(A_1(x, 1^m, E_{G_1(1^n)}(x)) = 1) - P(A_1(x, 1^m, E_{G_1(1^n)}(1^m)) = 1)$$
$$> (1 - 2^{-n})\frac{3}{8}n^{-c} - 2^{-n}$$
$$> \frac{3}{8}n^{-c} - \frac{1}{32}n^{-2c}.$$

- If $x \notin D_n$, yet there exists a ξ with

$$|p_{x,x}(\xi) - p_{x,1^m}(\xi)| \geq \frac{1}{8}n^{-2c}$$

(the event of these two conditions will be denoted by C_2), then with probability at least say $1 - 2^{-n}$, one of the following two alternatives happens: Either A_1 has terminated after its first step or the expression (estimator)

$$\widehat{p}_{x,x}(\xi^*) - \widehat{p}_{x,1^m}(\xi^*)$$

has the same sign as its "true" counterpart

$$p_{x,x}(\xi^*) - p_{x,1^m}(\xi^*).$$

It follows that

$$P(A_1(x, 1^m, E_{G_1(1^n)}(x)) = 1) - P(A_1(x, 1^m, E_{G_1(1^n)}(1^m)) = 1) > -2^{-n}$$
$$> -\frac{1}{32}n^{-2c}.$$

- In the remaining case (event C_3), independently of the fact if the estimator calculated in the first step of A_1 returns the correct result or not, one finds

$$P(A_1(x, 1^m, E_{G_1(1^n)}(x)) = 1) - P(A_1(x, 1^m, E_{G_1(1^n)}(1^m)) = 1)$$
$$\geq -\frac{1}{8}n^{-2c}.$$

Let $H(z,t)$ denote the event that A_1 yields 1 on input $(z, E_{G_1(1^n)}(t))$. We find

$$P(H(Z_n, X_n)) - P(H(Z_n, 1^m))$$
$$\geq \sum_{x \in \mathbb{B}^m} P(X_n = x) \cdot (P(H(x1^m, x)) - P(H(x1^m, 1^m)))$$

$$\geq P(C_1)(\frac{3}{8}n^{-c} - \frac{1}{32}n^{-2c}) - P(C_2)\frac{1}{32}n^{-2c} - P(C_3)\frac{1}{8}n^{-2c}$$
$$> \frac{1}{2}n^{-c}\frac{3}{8}n^{-c} - \frac{1}{32}n^{-2c} - \frac{1}{8}n^{-2c}$$
$$= \frac{1}{32}n^{-2c}.$$

So A_1 distinguished encryptions of the "halves" (i.e., X_n and 1^m) of the polynomial-time random variable Z_n.$\square$

Proposition 10.2. *Let the cryptosystem (G, E, D) be semantically secure as in Definition 10.3. Furthermore, suppose that f is polynomial-time computable, quantifiers are reversed in the sense that instead of the scheme*

$$\forall A \exists A' \forall \{X_n\} \forall h \forall f$$

("strongest" possible order) we have the "weakest" possible order

$$\forall A \forall \{X_n\} \forall h \forall f \exists A',$$

and that the conditional distribution of $f^{-1}(X_n)$ given $h(X_n)$ is a symmetric Bernoulli one. Then (G, E, D) satisfies the property of indistinguishable encryption.

Proof: The proof is of a similar nature to that of Proposition 10.1. Assume (G, E, D) has distinguishable encryptions and assume that there exists a polynomial-time random variable $\{T_n\}_{n\geq 1} = \{X_n Y_n Z_n\}_{n\geq 1}$, a probabilistic polynomial-time algorithm A and a positive constant c such that for infinitely many n, it holds that

$$|P(A(Z_n, E_{G_1(1^n)}(X_n)) = 1) - P(A(Z_n, E_{G_1(1^n)}(Y_n)) = 1)| > n^{-c}. \quad (10.5)$$

We may assume $|X_n| = |Z_n| = n$. Let $\{Q_n\}_{n\geq 1}$ be a (polynomial-time) random variable that takes the two possible values $0^n Z_n X_n Y_n$ and $1^n Z_n X_n Y_n$ both with probability one half. Let $f : B^{4n} \to B$ be the function that returns the first bit of every bitstring of length $4n$. On the other hand, define $h : B^{4n} \to B^{3n}$ as the function omitting the first n-block of its argument and, if this block was 1^n, interchanges the order of the other two last n-blocks of the argument. By this definition, the random variables $h(Q_n) = Z_n X_n Y_n$ and $f(Q_n)$ are independent. Also, f and h are computable in polynomial time and do not depend on A. Let us now construct a probabilistic polynomial-time algorithm A_2 guessing $f(Q_n)$ from $h(Q_n)$ and $E_{G_1(1^n)}(Q_n)$. Let δ be of the form $0^n wxv$ or $1^n wvx$. Again, A_2 will consist of effecting 2 steps on input $h(\delta) = wxv$ and $E_e(\delta)$.

- Ignoring $E_e(\delta)$, algorithm A_2 samples polynomially many times and calculates the difference estimator

$$\Delta(w, x, v) := P(A(w, E_{G_1(1^n)}(x)) = 1) - P(A(w, E_{G_1(1^n)}(v)) = 1) \quad (10.6)$$

such that this estimator differs from the true value by less than $\frac{1}{8}n^{-2c}$ with probability at least $1 - 2^{-n}$ say.

- W.l.o.g. we assume that the estimator calculated in (10.6) is positive. Then the algorithm A_2 gives the fourth n-block of $E_e(\delta)$ together with w as input to algorithm A und outputs 1 if A outputs 1 and 0 else.

It remains to do the performance analysis of A_2. Let H_n be the event that A_2 successfully guesses $f(Q_n)$ given $E_{G_1(1^n)}(Q_n)$ and $h(Q_n)$. Furthermore, L_n denotes the event that the estimator (10.6) has the correct sign. Then we obtain

$$
\begin{aligned}
&P(H_n | L_n, h(Q_n) = wxv) \\
&= P(f(Q_n) = 1)P(H_n | L_n, h(Q_n) = wxv, f(Q_n) = 1) \\
&\quad + P(f(Q_n) = 0)P(H_n | L_n, h(Q_n) = wxv, f(Q_n) = 0) \\
&= \frac{1}{2}P(A(w, E_{G_1(1^n)}(x)) = 1) + \frac{1}{2}P(A(w, E_{G_1(1^n)}(v)) = 0) \\
&= \frac{1}{2} + \frac{\Delta(w, x, v)}{2}.
\end{aligned}
$$

W.l.o.g. we may assume that $\Delta(w, x, v) \geq 0$. Now we split the situation into 3 possible cases:

- If $\Delta(w, x, v) \geq n^{-c}$ (denote this event by K_1), then with probability say $1 - 2^{-n}$ the estimator has the correct sign. So

$$
\begin{aligned}
P(H_n | h(Q_n) = wxv) &\geq P(\text{estimator correct})(\frac{1}{2} + \frac{1}{2}n^{-c}) \\
&> \frac{1}{2} + \frac{1}{2}n^{-c} - 2^{-n} \\
&> \frac{1}{2} + \frac{1}{4}n^{-c}.
\end{aligned}
$$

- If $\frac{1}{4}n^{-2c} \leq \Delta(w, x, v) < n^{-c}$ (this event will be called K_2), then here also with probability say $1 - 2^{-n}$, the sign of the estimator is the correct one and one calculates

$$
\begin{aligned}
P(H_n | h(Q_n) = wxv) &\geq P(\text{estimator correct})(\frac{1}{2} + \frac{1}{8}n^{-2c}) \\
&> \frac{1}{2}.
\end{aligned}
$$

- If the remaining event (which will be called K_3) holds (i.e., if $\Delta(w, x, v) < \frac{1}{4}n^{-2c}$), then

$$
P(H_n | h(Q_n) = wxv) \geq \frac{1}{2} - \frac{1}{8}n^{-2c}.
$$

Putting the three cases together, we thus obtain

$$P(H_n) = \sum_{j=1}^{3} P(K_j)P(H_n|K_j)$$

$$\geq n^{-c}(\frac{1}{2} + \frac{1}{4}n^{-c}) + (1 - n^{-c})(\frac{1}{2} - \frac{1}{8}n^{-2c})$$

$$> \frac{1}{2} + \frac{1}{8}n^{-2c}.$$

This means that the probability of success of the algorithm is significantly greater than $\frac{1}{2}$, which is a contradiction. $\square$

11 *Algorithmic Complexity

In this chapter, we present a "non-classical" definition of randomness that is of a quite different nature from the other criteria in the previous chapters. Namely, loosely speaking, a bitstring can be called "random", if the shortest program (in the sense of a Turing machine) for describing the string is the string itself. The length of this shortest program can be viewed as some sort of "algorithmic complexity" measure, which itself is of rather theoretical value, but one can show that it is indeed in "most" cases closely related to the linear complexity. So somewhat surprisingly, for "most" cases (in a measure-theoretic sense to be specified), the linear complexity seems to be a "universal" randomness criterion! (However, this definition does not apply to individual sequences!)

Definition 11.1. *The Turing-Kolmogorov-Chaitin complexity (TKC complexity for short) $\chi(x)$ of a bitstring $x \in \mathbb{B}^n$ of length n is the length of the shortest program for a universal Turing machine U that makes U simulate a Turing machine generating x.*

Unfortunately, we have

Proposition 11.1. *In general, the function χ is not computable.*

(Note that by the famous Church Thesis, computability by a Turing machine has turned out to be equivalent to all "reasonable" computability notions, e.g., to be a primitive recursive function.)
Proof of Proposition 11.1: Assume the contrary and let $K > 0$ be any constant and T_K a Turing machine that generates and inspects bitsequences in lexicographical order until a sequence x with $\chi(x) > K$ appears and then accepts this x. Denote by $\mu(T_K)$ the length of the program for U that makes U simulate T_K. Then we have

$$\mu(T_K) = O(\log K),$$

but on the other hand
$$\mu(T_K) < K.$$

So if K is chosen large enough, this yields a contradiction. $\square$

D. Neuenschwander: Prob. and Stat. Methods in Cryptology, LNCS 3028, pp. 135-138, 2004.
© Springer-Verlag Berlin Heidelberg 2004

However, one can estimate the average behavior of χ:

Proposition 11.2. *For $1 \leq k \leq n$ we have*

$$|\{x \in I\!B^n : \chi(x) \leq k\}| \leq 2^{k+1}.$$

Proof: This follows from the fact that among the 2^{k+1} Turing machines T with $\mu(T) \leq k$ there are at most 2^{k+1} sequence generators. $\square$

As a consequence, we obtain that "practically all sequences of moderately large length" have a TKC complexity close to the length of the sequence. In other words, a "truly random sequence has no shorter description than just the sequence itself".

Corollary 11.1. *If one assumes that the sequences x are uniformly distributed on $I\!B^n$, then*

$$P(\chi(x) > (1 - \varepsilon)n) > 1 - 2^{-\varepsilon n + 1} \quad (\varepsilon > 0).$$

Next we show that the TKC complexity $\chi(x)$ and the linear complexity $L(x)$ are asymptotically the same for "practically all" sequences $x \in I\!B^n$ (as $n \to \infty$).

Theorem 11.1. *Let ω be the Haar measure (uniform probability distribution) on $I\!B^N$. Then*

$$\frac{\chi(x^{(n)})}{L(x^{(n)})} \overset{\omega - a.s.}{\to} 1 \quad (n \to \infty).$$

(As usual, "a.s." means "almost surely", i.e. "$\omega(\ldots) = 1$".)

The proof needs a sequence of several lemmas.

Lemma 11.1. *For $t, u \in]0, n[$ we have*

$$|\{x \in I\!B^n : n + t < L(x)\}| \leq \frac{1}{3} 2^{n-t+1} - \frac{1}{3}, \tag{11.1}$$

$$|\{x \in I\!B^n : \chi(x) < n - u\}| \leq 2^{n-u+1}, \tag{11.2}$$

$$|\{x \in I\!B^n : L(x) \leq n + t, n - u \leq \chi(x)\}| \geq 2^n - \frac{1}{3} 2^{n-t+1} - 2^{n-u+1}. \tag{11.3}$$

Proof: Inequality (11.2) and the conclusion (11.1),(11.2)$\Longrightarrow$(11.3) are trivial. So it suffices to verify (11.1). By Corollary 7.1 we have indeed

$$|\{x \in I\!B^n : n + t < L(x)\}|$$

$$\leq \sum_{n+t<2i\leq 2n} 2^{2n-2i}$$

$$= \sum_{2\lfloor \frac{n+t}{2} \rfloor + 2 \leq 2i \leq 2n} 2^{2n-2i}$$

$$= 2^{2n-2\lfloor\frac{n+t}{2}\rfloor-2} \sum_{0\leq i\leq n-\lfloor\frac{n+t}{2}\rfloor-1} 2^{-2i}$$

$$= 2^{2n-2\lfloor\frac{n+t}{2}\rfloor-2} \frac{1 - 2^{-2(n-\lfloor\frac{n+t}{2}\rfloor)}}{3/4}$$

$$= \frac{1}{3}2^{2n-2\lfloor\frac{n+t}{2}\rfloor} - \frac{1}{3}$$

$$\leq \frac{1}{3}2^{n-t+1} - \frac{1}{3}.\square$$

Lemma 11.2.

$$|\{x \in I\!B^n : 0 \leq L(x) < n - t\}| \leq \frac{1}{3}2^{n-t+1} + \frac{1}{3}, \tag{11.4}$$

Proof:

$$|\{x \in I\!B^n : 0 \leq L(x) < n - t\}|$$

$$\leq 1 + \sum_{1\leq 2i<n-t} 2^{2i-1}$$

$$\leq 1 + \sum_{2\leq 2i<2\lfloor\frac{n-t}{2}\rfloor-2} 2^{2i-1}$$

$$= 1 + 2 \sum_{0\leq i\leq\lfloor\frac{n-t}{2}\rfloor-2} 2^{2i}$$

$$= 1 + 2\frac{2^{(\lfloor\frac{n-t}{2}\rfloor-1)} - 1}{3}$$

$$\leq \frac{1}{3}2^{n-t+1} + \frac{1}{3}.\square$$

Proposition 11.3. *For all* $\varepsilon \in]0, 1[$ *we have (under the uniform distribution of* x *on* $I\!B^n$*)*

$$P((1 - \varepsilon)L(x) \leq \chi(x)) \geq 1 - \frac{8}{3}2^{-\frac{\varepsilon}{2-\varepsilon}n}, \tag{11.5}$$

$$P((1 - \varepsilon)\chi(x) \leq L(x)) \geq 1 - \frac{1}{3}2^{-\varepsilon n+(1-\varepsilon)(1+\log n)+1} - \frac{1}{3}2^{-n}. \tag{11.6}$$

Proof: Put $t := \frac{\varepsilon}{2-\varepsilon}n$, so that $(1 - \varepsilon)(n + t) = n - t$. Now with the aid of Lemma 11.1 we get

$$P((1 - \varepsilon)L(x) \leq \chi(x))$$
$$\geq P(L(x) \leq n + t, (1 - \varepsilon)(n + t) \leq \chi(x))$$
$$= P(L(x) \leq n + t, n - t \leq \chi(x))$$
$$\geq 1 - \frac{1}{3}2^{-t+1} - 2^{-t+1}$$
$$= 1 - \frac{8}{3}2^{-\varepsilon n/(2-\varepsilon)},$$

which proves (11.5). For (11.6), put

$$t := \varepsilon n - (1 - \varepsilon)\lceil \log n \rceil,$$

so that

$$(1 - \varepsilon)(n + \lceil \log n \rceil) = n - t.$$

Now from Lemma 11.2

$$P((1 - \varepsilon)\chi(x) \leq L(x))$$
$$\geq P(\chi(x) \leq n + \lceil \log n \rceil, (1 - \varepsilon)(n + \lceil \log n \rceil) \leq L(x))$$
$$= P(n - t \leq L(x))$$
$$\geq 1 - \frac{1}{3}2^{-\varepsilon n + (1-\varepsilon)(1+\log n)+1} - \frac{1}{3}2^{-n}. \quad \square$$

Corollary 11.2. *For $\varepsilon \in]0, 1[$, we have (under the uniform distribution of x on $\mathbb{B}^{n+1}$)*

$$P((1 - \varepsilon)L(x^{(n)}) \leq \chi(x^{(n)}) \leq (1 + \varepsilon)L(x^{(n)})) \to 1 \quad (n \to \infty).$$

Proof of Theorem 11.1: For $x = (x_0, x_1, \ldots) \in \mathbb{B}^{\mathbb{N}}$ $(m \leq n)$, denote

$$x^{(m,n)} := (x_m, x_{m+1}, \ldots, x_n).$$

Define, for $k \in \mathbb{N}_0$ and $\varepsilon \in]0, 1[$, the (independent) events

$$A_{k,\varepsilon} := \{x \in \mathbb{B}^{\mathbb{N}} : (1 - \varepsilon)L(x^{(2^{k-1}, 2^k - 1)}) \leq \chi(x^{(2^{k-1}, 2^k - 1)})$$
$$\leq (1 + \varepsilon)L(x^{(2^{k-1}, 2^k - 1)})\}.$$

Then Corollary 11.2 yields

$$\sum_{k=1}^{\infty} \omega(A_{k,\varepsilon}) = \infty$$

and the assertion follows from the Borel-Cantelli Lemma. $\quad \square$

12 Birthday Paradox and Meet-in-the-Middle Attack

12.1 The Classical Birthday Attack

In this chapter, we will discuss the aspect of integrity, i.e., the danger that
Eve could change the message sent by Alice so that Bob does not notice that
the message he receives is now fraudulent.

The following so-called "birthday paradox" is well-known in probability theory: Suppose there are 23 persons in a group. Then the probability that there
exist two persons whose birthdays coincide is more than $1/2$. More generally,
consider a group of k persons and let n be the number of possible "birthdays"
(so in the above example $k = 23$ and $n = 365$). Let $q_{n,k}$ be the probability
that there exist no two persons with the same "birthday". One calculates

$$q_{n,k} = n^{-k} \prod_{i=0}^{k-1} (n - i)$$

$$= \prod_{i=1}^{k-1} (1 - \frac{i}{n})$$

$$\leq \prod_{i=1}^{k-1} e^{-i/n}$$

$$= e^{-k(k-1)/(2n)}.$$

So if

$$k \geq (1 + \sqrt{1 + 8n \log 2})/2, \tag{12.1}$$

we have $q_{n,k} \leq 1/2$.

Now we will introduce the notion of a so-called hash function. A hash function h is a function that maps bitstrings of arbitrary (but finite) length to
bitstrings of some fixed maximal length n. We will assume that for every
bitstring x, the image $h(x)$ is easy to compute, but that it is computationally
infeasible to find, for a given value y, an inverse image x such that $y = h(x)$.
The cryptologic application is that if x is the message that Alice wants to
transfer to Bob, then she in fact sends $h(x)$, which Eve can not invert. However, since x can be arbitrarily long, but $h(x)$ has maximal length n, there
must be collisions, i.e. bitstrings x, x' such that $x \neq x'$, but $h(x) = h(x')$. The

D. Neuenschwander: Prob. and Stat. Methods in Cryptology, LNCS 3028, pp. 139-144, 2004.
© Springer-Verlag Berlin Heidelberg 2004

hash function h is called strongly collision resistant if it is "infeasible" to find two different colliding bitstrings x, x' such that $h(x) = h(x')$. The so-called birthday attack is the attack to find such x, x' with not too small probability. By the preceding discussion, if one chooses, e.g., $k \geq (1 + \sqrt{1 + 8n \log 2})/2$ bitstrings $x_{(i)}$ ($1 \leq i \leq k$) and calculates their values $h(x_{(i)})$, then with probability more than $1/2$ there are two different $x_{(i_1)}, x_{(i_2)}$ such that

$$h(x_{(i_1)}) = h(x_{(i_2)}).$$

12.2 The Generalized Birthday Problem and Its Limit Distribution

In the following, we will consider the following variant of the birthday problem, which will be the key of the so-called meet-in-the-middle attack, that we will present in the next section. Consider a set E of n elements and draw two samples E_r and E_s of sizes r, resp. s, (with replacements) from it. What is the probability $P(n, r, s, i)$ that exactly i elements belong to both samples? Define

$$Q(n, r, k) := P(|E_r| = r - k)$$

(the probability of k coincidences in one sample (with replacements) of size r) and

$$H(n, r - k, s - \ell, i) := P(|E_r \cap E_s| = i \mid |E_r| = r - k, |E_s| = s - \ell)$$

(this is the probability that the intersection of the two samples of size $r - k$, resp. $s - \ell$, drawn without replacements contains i elements). Then we obtain

$$P(n, r, s, i) = P(\bigcup_{k=0}^{r-i} \bigcup_{\ell=0}^{s-i} \{|E_r| = r - k, |E_s| = s - \ell, |E_r \cap E_s| = i\})$$

$$= \sum_{k=0}^{r-i} \sum_{\ell=0}^{s-i} P(|E_r \cap E_s| = i, |E_r| = r - k \mid |E_s| = s - \ell) P(|E_r| = r - k, |E_s| = s - \ell)$$

$$= \sum_{k=0}^{r-i} \sum_{\ell=0}^{s-i} Q(n, r, k) H(n, r - k, s - \ell, i) Q(n, s, \ell). \qquad (12.2)$$

Standard combinatorial reasoning yields that $H(\ldots)$ is given by the hypergeometric distribution:

$$H(n, r, s, i) = \binom{r}{i} \binom{n - r}{s - i} / \binom{n}{s}. \qquad (12.3)$$

Now let us evaluate $Q(n, r, c)$. Clearly

$$Q(n, r, c) = 0 \qquad (r \geq n, c < r - n). \qquad (12.4)$$

In the other cases, observe that the c coincidences are drawn from a set of $r - c$ elements, which is equivalent to choosing from the r drawings α_1 ones to yield the first element, α_2 ones to yield the second element, etc., until one has chosen all $r - c$ elements. If we define

$$R := \{\alpha := (\alpha_1, \alpha_2, \ldots, \alpha_{r-c}) : \alpha_j \in \{1, 2, \ldots, c+1\}, \sum_{j=1}^{r-c} \alpha_j = r\},$$

then the number of ways each vector in R can be ordered is given by the multinomial coefficient $\binom{r}{\alpha}$. Hence

$$Q(n, r, c) = \frac{\binom{n}{r-c}}{n^r} \sum_{\alpha \in R} \binom{r}{\alpha}. \tag{12.5}$$

Let us now determine the limit of $P(n, r, s, i)$ (for $n, r, s \to \infty$ under suitable common behavior of n, r, and s). It turns out that it is given by a Poisson distribution as follows:

Theorem 12.1.

$$P(n, r, s, i) \to e^{-\nu} \frac{\nu^i}{i!} \qquad (n, r, s \to \infty, \frac{r^2}{2n} \to \lambda, \frac{s^2}{2n} \to \mu, \frac{rs}{n} \to \nu).$$

For the proof, we will separately (in the form of lemmas and corollaries) consider the asymptotic behavior of $H(\ldots)$ and $Q(\ldots)$. Then relation (12.2) will yield the result. The following limit theorem for the hypergeometric distribution is well known and easy to prove:

Lemma 12.1.

$$H(n, r, s, i) \to e^{-\nu} \frac{\nu^i}{i!} \qquad (n, r, s \to \infty, \frac{rs}{n} \to \nu).$$

Lemma 12.2.

$$Q(n, r, c) \to e^{-\lambda} \frac{\lambda^c}{c!} \qquad (n, r \to \infty, \frac{r^2}{2n} \to \lambda).$$

Proof: If $\alpha \in R$ with k components $\neq 1$ (evidently $k \leq c$), the summand $\binom{r}{\alpha}$ in the sum (12.5) occurs exactly $\binom{r-c}{k}$ times, hence, if we define R_k as the set of all non-decreasing sequences $\alpha = (\alpha_1, \alpha_2, \ldots, \alpha_k)$ of length k with elements in $\{2, 3 \ldots, c+1\}$ such that $\sum_{j=1}^{k} \alpha_j = c + k$, we can write

$$Q(n, r, c) = \frac{\binom{n}{r-c}}{n^r} \sum_{k=1}^{c} \frac{(r-c)!}{k!(r-c-k)!} \sum_{\alpha \in R_k} \binom{r}{\alpha}$$

$$\sim \frac{\binom{n}{r-c}}{n^r} (r-c)! \sum_{k=1}^{c} \frac{r^{c+k}}{k!} \sum_{\alpha \in R_k} (\prod_{i=1}^{k} \alpha_i!)^{-1}$$

$$= \frac{n!}{n^r (n-r-c)! \, 2^c c!} r^{2c} (1 + \frac{\gamma}{r})$$

$$\sim \frac{e^{-r^2/(2n)}}{n^c} \frac{r^{2c}}{2^c c!} (1 + \frac{\gamma}{r}),$$

with

$$\gamma := 2^c r^{1-c} c! \sum_{j=1}^{c-1} r^{-j}/j!.$$

So finally we find

$$Q(n,r,c) \sim e^{-r^2/(2n)} (\frac{r^2}{2n})^c /c!,$$

which yields the result. $\square$

For the determination of the limit of $P(\ldots)$, we study the behavior of $H(\ldots)$ in some more detail:

Lemma 12.3. *For $\frac{K}{N} < \frac{1}{2}$ we have the estimations*

$$\exp(-\frac{K^2}{2N} + \frac{K}{2N} - \frac{K^3}{3N^2}) \le \frac{N!}{N^K (N-K)!} \le \exp(-\frac{K^2}{2N} + \frac{K}{2N}).$$

This lemma yields the following two inequalities:

Corollary 12.1.

$$H(n,r,s,i) f_i(n,r,s,i,k,\ell) \le H(n,r-k,s-\ell,i),$$

where

$$f_i(n,r,s,i,k,\ell) := g(r,i,k) g(s,i,\ell) g(n,r,k) g(n,s,\ell) e^{\frac{(k+1)^2}{n-r-s}},$$

with

$$g(r,i,k) := e^{-\frac{k}{r} - \frac{k^2}{r-i}} (1 - \frac{i}{r})^k.$$

Corollary 12.2.

$$H(n,r-k,s-\ell,i) \le H(n,r,s,i) \tilde{f}_s(n,r,s,i,k,\ell),$$

where

$$\tilde{f}_s(n,r,s,i,k,\ell) := \tilde{g}(n,r,i,k) \tilde{g}(n,s,i,\ell) e^{\frac{(k+1)^2}{n-r-s} + 2(k+1)\frac{r+s}{n}},$$

with

$$\tilde{g}(n,r,i,k) := e^{\frac{k^2}{r-i} + \frac{k}{r-i} + \frac{k}{n-r}}.$$

With these two corollaries, we obtain the limits

Corollary 12.3.

$$f_i(n,r,s,i,k,\ell), \tilde{f}_s(n,r,s,i,k,\ell) \to 1 \quad (n,r,s \to \infty, \frac{r^2}{2n} \to \lambda, \frac{s^2}{2n} \to \mu, \frac{rs}{n} \to \nu).$$

Now we may proceed to the proof of Theorem 12.1. For fixed α, β, we have, since all occurring terms are positive,

$$P(n,r,s,i) \geq \sum_{k=0}^{\alpha} \sum_{\ell=0}^{\beta} Q(n,r,k)H(n,r-k,s-\ell,i)Q(n,s,\ell)$$
$$\geq H(n,r,s,i)f_i(n,r,s,i,\alpha,\beta)$$
$$\cdot \sum_{k=0}^{\alpha} Q(n,r,k) \sum_{\ell=0}^{\beta} Q(n,s,\ell). \tag{12.6}$$

Taking the limit of the first, resp. second, sum in the last member of inequality (12.6) yields the probability distribution functions of a Poisson distribution with parameter λ, resp. μ, evaluated at α, resp. β.

On the other hand, since $H(n,r-k,s-\ell,i) \leq 1$ ($k \geq \alpha$ or $\ell \geq \beta$) and $H(n,r-k,s-\ell,i) \leq H(n,r,s,i)$ (else), we get

$$P(n,r,s,i) \leq H(n,r,s,i)\tilde{f}_s(n,r,s,i,\alpha,\beta) \sum_{k=0}^{\alpha} \sum_{\ell=0}^{\beta} Q(n,r,k)Q(n,s,\ell)$$
$$+ \sum_{k=\alpha+1}^{r-i} Q(n,r,k) \sum_{\ell=0}^{\beta} Q(n,s,\ell)$$
$$+ \sum_{k=0}^{\alpha} Q(n,r,k) \sum_{\ell=\beta+1}^{s-i} Q(n,s,\ell)$$
$$+ \sum_{k=\alpha+1}^{r-i} Q(n,r,k) \sum_{\ell=\beta+1}^{s-i} Q(n,s,\ell).$$

On the right-hand side, again $\tilde{f}_s(n,r,s,i,\alpha,\beta) \leq 1$, whereas the sums tend to the corresponding Poisson distribution functions $F_{.}$, resp. to $1 - F_{.}$, more precisely:

$$\lim P(n,r,s,i) \leq \lim H(n,r,s,i)F_\lambda(\alpha)F_\mu(\beta)$$
$$+(1 - F_\lambda(\alpha))F_\mu(\beta) + F_\lambda(\alpha)(1 - F_\mu(\beta))$$
$$+(1 - F_\lambda(\alpha))(1 - F_\mu(\beta)). \tag{12.7}$$

Now letting tend $\alpha, \beta \to \infty$ yields Theorem 12.1.

12.3 The Meet-in-the-Middle Attack

Here, we consider the so-called Rabin scheme, which is given as follows: Let the plaintexts x consist of n blocks $x_{(1)}, x_{(2)}, \ldots, x_{(n)} \in \mathbb{B}^{56}$. Similarly, the ciphertext y will be written as

$$y =: (y_{((1))}, y_{(2)}, \ldots, y_{(n)}) \in (I\!B^{56})^n.$$

Denote by $E_k(x)$ the DES encrypting of the plaintext x with key k and $D_k(y)$ the DES deciphering of the ciphertext y with key k (for the description of DES see any general manual on modern cryptology). Then the hash values $h_1, h_2, \ldots, h_n$ are defined as

$$h_j := E_{x_{(j)}}(h_{j-1}) \quad (1 \le j \le n),$$

where h_0 is some uniformly distributed random element of $I\!B^{56}$. We will write

$$E_x(h) := E_{x_{(n)}}(E_{x_{(n-1)}}(\ldots(E_{x_{(1)}}(h))\ldots)),$$

$$D_y(h) := E_{y_{(1)}}(E_{y_{(2)}}(\ldots(E_{y_{(n)}}(h))\ldots)).$$

Now for mounting the so-called meet-in-the-middle attack, Eve first generates 2^{32} messages $x_{[\ell]}$ and $x_{[r]}$, and calculates their values $h_\ell(= E_{x_{[\ell]}}(h_0))$ and $h_r(= D_{x_{[r]}}(h_{2^{n_r}}))$ (where n_r denotes the number of 56-bit blocks of $x_{[r]}$). Now she sorts the lists of all values h_ℓ, resp. of all values h_r, (recall that sorting is "fast" in the sense that sorting a list of n elements requires $O(n \log n)$ operations). If one supposes that E encrypts "randomly enough", this can be considered (before ordering) as two random samples of 2^{32} drawings with replacements of a total population of 2^{64} elements. So by Theorem 12.1, during the sorting, a coincidence (i.e., a case where there exist some ℓ_0, r_0 such that $h_{\ell_0} = h_{r_0}$) occurs with probability at least about $1 - e^{-1}$. Now put

$$x := (x_{[\ell_0]}, x_{[r_0]}).$$

Then we get

$$E_x(h_0) = E_{x_{[\ell_0]}}(h_{\ell_0}) = E_{x_{[r_0]}}(h_{r_0}) = h_{n_{r_0}}.$$

So h_0 and $h_{n_{r_0}}$ are, as one says, "linked up" (or "joined up") by x. Hence Eve can construct a fraudulent message x'.

13 Quantum Cryptography

In this final short chapter, we will present the fundamental idea of quantum cryptography. This is not the same thing as quantum computing treated in Chapter 3: There, quantum computers are used to cryptanalyze classical cryptosystems.

The most fundamental method of quantum cryptography can be demonstrated by the following example:

- Alice sends Bob a string of photon pulses. She polarizes every photon (randomly) in one of 4 possible directions: horizontal, vertical, left-diagonal, or right-diagonal, for example,

$$||/--|-/$$

- Bob is in possession of a polarization detector, which he can set to measure the rectilinear or diagonal polarization. For this, he can, for example, use a calcium carbonate crystal. Since in this material electrons are bound with different strengths in different directions, a photon passing through the crystal "feels" a different electromagnetic force depending on the orientation of the electric field relative to the polarization axis in the crystal. Bob can not measure both types of polarization, since in quantum mechanics measuring the one destroys the possibility of measuring the other (see also Chapter 3 for more details). If he sets his detector to measure rectilinear polarization and if Alice polarized her photon really as "horizontal" ($-$) or "vertical" ($|$), then Bob will learn how Alice polarized her photon. The same is true if Alice polarized as "\" or "/" and Bob measures diagonal polarization. However, if he sets the detector to measure rectilinear polarization and if Alice polarized diagonally, Bob will obtain a random measurement and, what is more, he will not know the difference. So Bob will set his detector at random, e.g.,

$$drrdddrdrr$$

(where "r" means "rectilinear" and "d" stands for "diagonal"). In our example, he could, for example, obtain the result

$$/|-/-|$$

D. Neuenschwander: Prob. and Stat. Methods in Cryptology, LNCS 3028, pp. 145-146, 2004.
© Springer-Verlag Berlin Heidelberg 2004

- Bob tells Alice (over an insecure channel) his detector settings.
- Alice tells Bob which settings (rectilinear or diagonal) were correct. Here, for example, the detector was correctly set for the photon pulses numbers 2, 6, 7, and 9.
- Alice and Bob keep only those polarizations that were correctly measured, so here

$$* \,|\, * \;\; * \;\; * \;\; * \;-\; *$$

These correctly measured polarizations can be used as a message (or a key) in the form of a bitstring by a prearranged code.

Since Bob will guess correctly in half of the cases, in order to generate n bits one has to use about $2n$ photon pulses. The important feature of quantum cryptography is that Eve can really not eavesdrop. Like Bob, she has to guess which type of polarization (rectilinear or diagonal) she has to measure, and she will be wrong in half of the cases. But then the polarization of the photons is changed, and Alice and Bob after comparing their bitstrings at the end, will find discrepancies, which shows them that there has been an attack by Eve. So they will just not use these bits and create new ones. By doing enough comparisons, they can get arbitrarily good security against an eavesdropping by Eve.

For more precise and further information on quantum cryptography, see, for example, Clearwater, Williams (1998), Chapter 8 or Hungerbühler, Struwe (2003).

Bibliographical Remarks

Section 1.1 (classical Vigenère Cipher) is taken from Beutelspacher (1993), whereas the remarks on perfect secrecy in Section 1.2 are based on Buchmann (2001), 4.4. For the counterexample and Theorem 1.2 on cascade ciphers see Maurer, Massey (1993).

The RSA system in Chapter 2 can be found in almost every introductory book on cryptology. Specifically, the treatment of the primality tests in Sections 2.2 (Soloway-Strassen Test) and 2.3 (Rabin Test) is taken from the book of Kranakis (1986). The subject of Section 2.4 (bit security, see also the key word "hard bits" in the literature) is somewhat more involved and the details (see Delfs, Knebl (2002), 7) can be omitted at first reading. Section 2.5 is about hardware implementation problems (timing attacks, see Boneh (1999)). Section 2.6 is a short glimpse into the huge and important subject of Zero Knowledge, see Schneier (1996), p. 548f.

Chapter 3 is about quantum probability, which behaves quite differently from classical probability. First, we give a short introduction to quantum computing (Clearwater, Williams (1998), Gruska (1999)). The preparatory Section 3.3 on continued fractions can be found, e.g., in the classical book of Perron (1954) or in any other introductory text on continued fractions. Shor's factorization algorithm (Section 3.4) is described as in Clearwater, Williams (1998).

The considerations about physical ("genuine") random-number generators in Sections 4.1 and 4.2 is an abbreviated version of the article Neuenschwander, Zeuner (1993). Section 4.3 proves a result of Näslund, Russell (2001) about the possibility to construct unbiased random bits from several biased ones with common rational bias by adding them mod.2.

In contrast to Chapter 4, in Chapter 5 we discuss pseudo-random-number generators as the linear feedback shift registers (LFSR) (especially the question in which cases they generate so-called pseudo-noise (PN) sequences (see van Tilborg (1988)). The so-called shrinking generator has been introduced by Coppersmith et al. (1994), whereas its variant, the self-shrinking generator, is due to Meier and Staffelbach (1995) (see also Blackburn (1999) proving the conjecture about its maximum linear complexity uttered by Meier and Staffelbach (1995)). For the notion of perfect pseudorandomness in Section 5.3, see Schrift, Shamir (1993). Section 5.4 has been taken from Massey

(1997), whereas for the correlation immunity in Section 5.5 see mainly Rueppel (1986) and Siegenthaler (1984). The quadratic congruential generator (Section 5.6) has been analyzed in Brands, Gill (1996) (see also the preceeding article Brands, Gill (1995)).

The information theory primer of Chapter 6 is mainly standard material. For Section 6.1, we have used the course notes of Carnal (1993). For Section 6.2, see mainly Massey (1997). Section 6.3 describes a new and relatively nonstandard approach to information theory (marginal guesswork) due to Pliam (2000).

Chapter 7 mainly treats the tests used to evaluate the AES ("Advanced Encryption Standard"), which is the successor of the DES ("Data Encryption Standard"). For the development of an AES, the National Institute for Standards and Technolgy (NIST) invited the worldwide community of cryptologists to a competition, which was finally won by Daemen and Rijmen with their algorithm RIJNDAEL (see Daemen, Rijmen (2002), see also Zürcher (2003)). For the tests themselves we refer, e.g., to Rukhin et al. (2000). The origin and certain details about the tests suggested by Rukhin can be found in the references cited in the main text. Section 7.1 is based mainly on the "Handbook of Cryptology" of the Swiss Army (1981). Proposition 7.1 can be found in Baron, Rukhin (1999). The "Approximate Entropy Test" in Section 7.7 has also been described by Rukhin (2000a,b). Proofs of the Berlekamp-Massey Algorithm (Section 7.11) can be found in many texts. Our treatment is that of the above-cited Handbook of Cryptology of the Swiss Army (1981), A.4. The statistics of the linear complexity has been given in Rueppel (1986) and Rukhin (2000b).

Chapter 8 is about the distribution of keys in the Diffie-Hellman system (Massey, Waldvogel (1993) and the literature cited therein). The Prime Number Theorem cited at the end of Chapter 8 can be found in any primer about analytic number theory.

Differential cryptanalysis, which is the subject of Chapter 9, and moreover its counterpart "linear cryptanalysis" (attributed to Matsui (1994)) has become very popular in recent years. The question to whom to attribute it is difficult, some first steps were probably made in Great Britain already several decades ago, but the developers were not allowed to publish it. Good explanations about different aspects can be found in Massey (1997), on which also our treatment is mainly based. Section 9.2 adresses the question of the distribution of characteristics as has been described in Neuenschwander (2002), a generalization of certain of O'Connor's (1995) results for bitstrings to sequences of elements of an arbitrary residue ring. Closely related is also the paper Hawkes, O'Connor (1999).

Chapter 10 is about the notion of semantic security (Goldreich (1993)).

The algorithmic complexity (or Turing-Kolmogorov-Chaitin complexity) discussed in Chapter 11 is of rather theoretical interest (see Beth, Dai (1990)). Chapter 12 addresses consequences of the well-known "birthday" paradox in probability theory for cryptology (especially hash functions). In many cryptology texts, one can find the keyword "birthday attack". In particular, Sections 12.2 and 12.3 are based on Campana et al. (1988).

Finally, Chapter 13 is an informal standard short introduction to quantum cryptography. A more sophisticated treatment of it can, e.g., be found in Clearwater, Williams (1998). See also Hungerbühler, Struwe (2003).

References

1. Agrawal, M., Kayal, N, Saxena, N.[1] (2003). *PRIMES is in P.* Manuscript Department of Computer Science and Engineering, Indian Institute of Technology Kanpur. Available on the Internet under `www.cse.iitk.ac.in/news/primality.html`.
2. Aldous, D., Shields, P. (1988). A Diffusion Limit for a Class of Randomly Growing Binary Trees. *Prob. Theory Rel. Fields* **79**, 509-542.
3. Banks, D., Dray, J., Leigh, S., Levenson, M., Nechvatal, J., Rukhin, A. L., Smid, M., Soto, J., Vangel, M., Vo, S. (2000). *A Statistical Test Suite for the Validation of Cryptographic Random Number Generators.* Special NIST Publication, NIST, Gaithersburg MD.
4. Barbour, A. D., Holst, L., Janson, S. (1992). *Poisson Approximation.* Clarendon Press, Oxford.
5. Baron, M., Rukhin, A. L. (1999). Distribution of the Number of Visits of a Random Walk. *Comm. Statist. - Stochastic Models* **15(3)**, 593-597.
6. Barton, D. E., David, F. N. (1962). *Combinatorial Chance.* Hafner, New York.
7. Bernstein, D. (2002). *An Exposition of the Agrawal-Kayal-Saxena Primality-Proving Theorem.* Manuscript. Available on the Internet under `cr.yp.to/papers.html♯aks`.
8. Beth, T., Dai, Z.-D. (1990). On the Complexity of Pseudo-Random Sequences - or: If You Can Describe a Sequence It Can't Be Random. In: Quisquater, J.-J., Vandwalle, J. (ed.). Adv. Crypt. EUROCRYPT '89. *Lecture Notes in Computer Science* **434**. Springer, Berlin, 533-543.
9. Beutelspacher, A. (1993). *Kryptologie.* 3. Auflage. Vieweg, Braunschweig.
10. Biham, E., Shamir, A. (1991). Differential Cryptanalysis of DES-like Cryptosystems. *J. Cryptology* **4(1)**,3-72.
11. Billingsley, P. (1956). Asymptotic Distributions of Two Goodness of Fit Criteria. *Ann. Math. Statist.* **27**, 1123-1129.
12. Blackburn, S. R. (1999). The Linear Complexity of the Self-Shrinking Generator. *IEEE Trans. Inf. Theory* **45(6)**, 2073-2076.
13. Blum, M., Micali, S. (1984). How to Generate Cryptographically Strong Sequences of Pseudo-Random Bits. *SIAM J. Computing* **13(4)**, 850-864.
14. Boneh, D. (1999). Twenty Years of Attacks on the RSA Cryptosystem. *Notices Am. Math. Soc.* **46(2)**, 203-213.
15. Boneh, D., Venkatesan, R. (1998). Breaking RSA May Not Be Equivalent to Factoring. In: Nyberg, K. (ed.). Adv. Crypt. EUROCRYPT'98. *Lecture Notes in Computer Science* **1403**. Springer, Berlin, 59-71.

---

[1] In this bibliography, for co-authored literature, we always put the names of the authors in alphabetic order.

16. de Bonis, A., de Santis, A (2001). New Results on the Randomness of Visual Cryptography Schemes. In: Lam, K.-Y. et al. (ed.). *Cryptography and Computational Number Theory.* Prog. Comput. Sci. Appl. Log. **20**, Birkhäuser, Basel, 187-201.

17. Bornemann, F. (2002). Primes in P: Ein Durchbruch für "Jedermann". *DMV Mittelungen* **4-2002**, 14-21.

18. Brands, S, Gill, R. (1995). Cryptography, Statistics, and Psudorandomness I. *Prob. Math. Stat.* **15**, 101-114.

19. Brands, S, Gill, R. (1996). Cryptography, Statistics, and Psudorandomness II. *Prob. Math. Stat.* **16(1)**, 1-17.

20. Brynielsson, L. (1989). A Short Proof of the Xiao-Massey Lemma. *IEEE Trans. Inf. Theory* **35(6)**, 1344.

21. Buchmann, J. A. (2001). *Introduction to Cryptography.* Springer, Berlin.

22. Campana, M., Cohen, R., Girault, M. (1988). A Generalized Birthday Attack. In: Günther, C. G. (ed.). Adv. Crypt. EUROCRYPT'88. *Lecture Notes in Computer Science* **330**. Springer, Berlin, 129-156.

23. Canetti, R., Friedlander, J., Shparlinski, I. (1999). On certain Exponential Sums and the Distribution of Diffie-Hellman Triples. *J. London Math. Soc. (2)* **59**, 799-812.

24. Carnal, H. (1993). *Informationstheorie.* Course Notes, University of Bern (CH).

25. Chepyzhov, V., Smeets, B. (1991). On a Fast Correlation Attack on certain Stream Ciphers. In: Davies, D. W. (ed.). EUROCRYPT'91. *Lecture Notes in Computer Science* **547**. Springer, Berlin, 176-185.

26. Clearwater, S. H., Williams, C. P. (1998). *Explorations in Quantum Computing.* Springer, Berlin.

27. Coppersmith, D., Krawczyk, H., Mansour, Y. (1994). The Shrinking Generator. In: Stinson, D. R. (ed.). Adv. Crypt. CRYPTO'93. *Lecture Notes in Computer Science* **773**. Springer, Berlin, 22-39.

28. Coron, J. S., Naccache, D. (1999). An Accurate Evalutation of Maurer's Universal Test. In: Selected Areas in Cryptography. *Lecture Notes in Computer Science* **1556**. Springer, Berlin, 57-71.

29. Daemen, J., Rijmen, V. (2002). The Design of Rijndael. AES - the Advanced Encryption Standard. Springer, Berlin.

30. Darmgård, I. B., Landrock, P., Pomerance, C. (1993). Average Case Error Estimates for the Strong Probable Prime Test. *Math. of Computation* **61(203)**, 177-194.

31. Delfs, H., Knebl, H. (2002). *Introduction to Cryptography.* Springer, Berlin.

32. Diffie, W., Hellman, M. (1976). New Directions in Cryptography. *IEEE Trans. Inf. Theory* **IT 22**, 644-654.

33. Feller, W. (1968). *An Introduction to Probability Theory and its Applications.* 3rd edn. Wiley, New York.

34. Goldreich, O. (1993). A Uniform-Complexity Treatment of Encryption and Zero-Knowledge. *J. Cryptology* **6(1)**, 21-53.

35. Gonzales Vasco, M. I., Shparlinski, I. E. (2001)- On the Security of Diffie-Hellman Bits. In: Lam, K.-Y. et al. (ed.). *Cryptography and Computational Number Theory.* Prog. Comput. Sci. Appl. Log. **20**, Birkhäuser, Basel, 257-268.

36. Good, I. J. (1953). The Serial Test for Sampling Numbers and Other Tests for Randomness. *Proc. Cam. Phil. Soc.* **49**, 276-284.

37. Good, I. J. (1957). On the Serial Test for Random Sequence. *Ann. Math. Statist.* **23**, 262-264.

38. Gruska, J. (1999). *Quantum Computing*. McGraw-Hill, London.
39. Guibas, L. J., Odlyzko, A. M. (1981). String Overlaps, Pattern Matching, and Nontransitive Games. *J. Combin. Theory* **A**, **30**, 183-208.
40. Hardy, G. H., Wright, E. M. (1960). *An Introduction to the Theory of Numbers*, 4th edn. Oxford University Press.
41. Hawkes, P., O'Connor, L. (1999). XOR and non-XOR Differential Probabilities. In: Stern, J. (ed.). Adv. Crypt. EUROCRYPT'99. *Lecture Notes in Computer Science* **1592**. Springer, Berlin, 272-285.
42. Hungerbühler, N., Struwe, M. (2003). A One-way Function from Thermodynamics and Applications to Cryptography. *Elem. Math.* **58**, 49-64.
43. Inoue, H., Kumahora, H., Yoshizawa, Y., Ichimura, M., Miyatake, O. (1983). Random Numbers Generated by a Physical Device. *Appl. Statist.* . **32(2)**, 115-120.
44. Kalouptsidis, N., Kolokotronis, N. (2003). On the Linear Complexity of Nonlinearly Filtered PN-sequences. *IEEE Trans. Inf. Theory* **49(11)**, 3047-3059.
45. Kirschenhofer, P., Prodinger, H., Szpankowski, W. (1994). Digital Search Trees Again Revisited: The Internal Path Length Perspective. *SIAM J. Computation* **23**, 598-616.
46. Koblitz, N. (1999). *Algebraic Aspects of Cryptography*. Springer, Berlin.
47. Kovalenko, I. N. (1972). Distribution of the Linear Rank of a Random Matrix. *Theory Prob. Appl.* **17**, 342-346.
48. Kranakis, E. (1986). *Primality and Cryptography*. Wiley, New York.
49. Lidl, R., Niederreiter, H. (1986). *Introduction to Finite Fields and their Applications*. Cambridge University Press, Cambridge.
50. Marsaglia, G. (1968). Random Numbers Fall Mainly in the Plane. *Proc. Nat. Acad. Sci.* **61**, 25-28.
51. Massey, J. L. (1997). *Cryptography: Fundamentals and Applications*. Advanced Technology Seminars, Zürich.
52. Massey, J. L., Maurer, U. M. (1993). Cascade Ciphers: The Importance of Being First. *J. Cryptology* **6(1)**, 55-61.
53. Massey, J. L., Waldvogel, C. P. (1993). The Probability Distribution of the Diffie-Hellman Key. In: Seberry, J. (ed.). Adv. Crypt. AUSCRYPT'92. *Lecture Notes in Computer Science* **718**. Springer, Berlin, 492-504.
54. Matsui, M. (1994). Linear Cryptanalysis Method for DES Cipher. In: Helleseth, T. (ed.). Adv. Crypt. EUROCRYPT'93. *Lecture Notes in Computer Science* **765**. Springer, Berlin, 386-397.
55. Maurer, U. M. (1992). A Universal Statistical Test for Random Bit Generators. *J. Cryptology* **5**, 89-105.
56. Meier, W. ,Staffelbach, O. (1989). Fast Correlation Attacks on certain Stream Ciphers. *J. Cryptology* **1(3)**, 159-176.
57. Meier, W., Staffelbach, O. (1991). Correlation Properties of Combiners with Memory in Stream Ciphers. In: Damgard, I. B. (ed.). Adv. Crypt. EUROCR¿PT'90. *Lecture Notes in Computer Scinece* **473**. Springer, Berlin, 204-213.
58. Meier, W., Staffelbach, O. (1992). Correlation Properties of Combiners with Memory in Stream Ciphers. *J. Cryptology* **5(1)**, 67-86.
59. Meier, W., Staffelbach, O. (1995). The Self-Shrinking Generator. In: De Santis, A. (ed.). Adv. Crypt. EUROCRYPT '94. *Lecture Notes in Computer Science* **950**, Springer, Berlin, 205-214.
60. Mitra, S. K., Rao, C. R. (1971). *Generalized Inverse of Matrices and Its Applications*. Wiley, New York.

61. Müller, S. (2003). A Probable Prime Test with very high Confidence for $n \equiv 3$ mod. 4, *J. Cryptology* **16(2)**, 117-139.
62. Näslund, M., Russell, A. (2001). Achieving Optimal Fairness from Biased Coinflips. In: Lam, K.-Y. et al. (ed.). *Cryptography and Computational Number Theory*. Birkhäuser, Basel, 303-330.
63. Neuenschwander, D. (2002). A Limit Theorem in Cryptology: The Asymptotic Distribution of Additive Characteristics of Random Permutations of $(\mathbb{Z}/q\mathbb{Z})$. In: Berkes, I. et al. (ed.). *Limit Theofems in Probability and Statistics II*. Proceedings of the 1999 Balatonlelle Conference. Budapest, 437-442.
64. Neuenschwander, D., Zeuner, H. M. (2003). Generating Random Numbers of Prescribed Distribution Using Physical Sources. *Stat. and Comp.* **13(1)**, 5-11.
65. von Neumann, J. (1963). Various Techniques Used in Connection with Random Digits. In: *von Neumann's Collected Works Vol. 5*. Pegamont Press, Elmsford NY, 768-770.
66. Nisley, E. (1990). Basic Radioactive Randoms. *Circuit Cellar Ink*, 58-68.
67. O'Connor, L. (1995). On the Distribution of Characteristics in Bijective Mappings. *J. Crpytology* **8**, 67-86.
68. Odlyzko, A. M. (2001). The 10^{22}-nd Zero of the Riemann Zeta Function. In: Lapidus, M. L. et al. (ed.). *Dynamical, spectral, and arithmetic zeta functions*. *Contemp. Math.* **290**, 139-144.
69. Okeya, K., Sakurai, K. (2000). Power Analysis Breaks Elliptic Curve Cryptosystems even Secure against the Timing Attack. In: Boy, B., Okamot, E. (ed.). Progr. Crypt. INDOCRYPT 2000. *Lecture Notes in Computer Science* **1977**. Springer, Berlin,178-190.
70. Perron, O. (1954). *Die Lehre von den Kettenbrüchen*. Teubner, Stuttgart.
71. Pliam, J. O. (2000). On the Incomparability of Entropy and Marginal Guesswork in Brute-Force Attacks. In: Boy, B., Okamot, E. (ed.). Progr. Crypt. INDOCRYPT 2000. *Lecture Notes in Computer Science* **1977**. Springer, Berlin, 67-79.
72. Reeds, J. A., Sloane, N. J. A. (1985). Shift-register Synthesis (modulo m). *SIAM J. Comput.* **14(3)**, 505-513.
73. Révész. P. (1990). *Random Walk in Random and Non-Random Environments*. World Scientific, Singapore.
74. Richter, M. (1993). PURAN 2: Ein Zufallsgenerator zur Erzeugung von quasiidelaen Zufallszahlen aus elektronischem Rauschen. *Informatik aktuell* **41**, 49-62.
75. Rueppel, R. A. (1986). *Analysis and Design of Stream Ciphers*. Springer, Berlin.
76. Rukhin, A. L. (2000a). Approximate Entropy for Testing Randomness. *J. Appl. Prob.* **37**, 88-100.
77. Rukhin, A. L. (2000b). Testing Randomness: A Suite of Statistical Procdures. *Theory Probab. Appl.* **45(1)**, 111-132.
78. Schindler, W. (2000). A Timing Attack Against RSA with the Chinese Remainder Theorem. In: Koc, C. K., Paar, C. (ed.). Cryptographic Hardware and Embedded Systems. *Lecture Notes in Computer Science* **1965**. Springer, Berlin, 109-124.
79. Schindler, W. (2002a). Optimized Timing Attacks Against Public Key Cryptosystems. *Statistics and Decisions* **20**, 191-210.
80. Schindler, W. (2002b). A Combined Timing and Power Attack. In: Paillier, P., Naccache, D. (ed.). Public Key Cryptography 2002. *Lecture Notes in Computer Science* **2274**. Springer, Berlin, 263-279.

81. Schindler, W., Walter, C. (2003). More Detail for a Combined Timing and Power Attack against Implementations of RSA. In: Paterson, K. G. (ed.). Cryptography and Coding - IMA 2003. *Lecture Notes in Computer Science* **2898**. Springer, Berlin, 245-263.

82. Schneier, B. (1996). *Applied Cryptography*. Wiley, New York.

83. Schrift, A. W., Shamir, A. (1993). Universal Tests for Nonuniform Distributions. *J. Cryptology* **6**, 119-133.

84. Seifert, J.-P. (2001). Using Fewer Qubits in Shor's Factorization Algorithm Via Simultaneous Diophantine Approximation. In: Naccache, D. (ed.). Topics in Cryptology. CT-RSA 2001. *Lecture Notes in Computer Science* **2020**. Springer, Berlin, 319-327.

85. Siegenthaler, T. (1984). Correlation Immunity of Nonlinear Combining Functions for Cryptographic Applications. *IEEE Trans. Inf. Theory* **IT-30(5)**, 776-780.

86. Swiss Army (1981). *Kryptologen-Handbuch*. Bern.

87. van Tilborg, H. C. A. (1988). *An Introduction to Cryptology*. Kluwer, Boston.

88. Walker, H. (1996). *HotBits: Genuine Random Numbers Generated by Radioactive Decay*. Fourmilab, http://www.fourmilab.ch/hotbits.

89. Walther, U. (1999). *Verschlüsselungssysteme auf Basis endlicher Geometrien*. Ph. D. Thesis University of Giessen. Mittelungen aus dem Math. Seminar Giessen, Heft **236**, Selbstverlag des Math. Instituts.

90. Williams, H. C. (1980). A Modification of the RSA Public-Key Encryption Procedure. *IEEE Trans. Inf. Theory* **26**, 726-729.

91. Zürcher, M. (2003). *Security of the Advanced Encryption Standard*. Diploma Thesis University of Bern (CH).

Index